St. George Jackson Mivart

**Types of Animal Life**

St. George Jackson Mivart

**Types of Animal Life**

ISBN/EAN: 9783337095390

Printed in Europe, USA, Canada, Australia, Japan

Cover: Foto ©berggeist007 / pixelio.de

More available books at **www.hansebooks.com**

# TYPES

OF

# ANIMAL LIFE

BY

ST. GEORGE MIVART, F.R.S.
AUTHOR OF
"ESSAYS AND CRITICISMS," ETC.

*WITH ILLUSTRATIONS*

LONDON
JAMES R. OSGOOD, M<sup>c</sup>ILVAINE & CO.
45 ALBEMARLE STREET, W.
1893

# CONTENTS

| | PAGE |
|---|---|
| I. MONKEYS | 1 |
| II. THE OPOSSUM | 36 |
| III. THE TURKEY | 66 |
| IV. THE BULLFROG | 96 |
| V. THE RATTLESNAKE | 122 |
| VI. THE SEROTINE, OR CAROLINA BAT | 150 |
| VII. THE AMERICAN BISON | 177 |
| VIII. THE RACOON | 211 |
| IX. THE SLOTH | 246 |
| X. THE SEA-LION | 275 |
| XI. WHALES AND MERMAIDS | 303 |
| XII. THE OTHER BEASTS | 336 |

# LIST OF ILLUSTRATIONS

| | PAGE | | PAGE |
|---|---|---|---|
| The Siamang Gibbon | 13 | The Proteus | 110 |
| The Proboscis Monkey | 16 | The Amblystoma | 111 |
| The White-nosed Monkey | 19 | The Axolotl | 112 |
| The Wanderoo | 21 | The Cœcilia | 114 |
| The Black Macaque | 22 | The Common Rattlesnake | 123 |
| The Chacma | 23 | The Rhinoceros Viper | 132 |
| The Long-haired Spider Monkey | 25 | The India Cobra | 136 |
| The Bald-headed Saki | 29 | The Two-coloured Sea-snake | 139 |
| The Squirrel Monkey | 31 | The Boa-Constrictor | 140 |
| The Virginian Opossum | 37 | The Carolina Bat | 152 |
| The Vulpine Phalanger | 45 | The Long-Eared Bat | 160 |
| The Tasmanian Wolf | 47 | The Megaderma Lyra | 163 |
| The Koala | 49 | The Vampire | 167 |
| The Echidna | 50 | The Kalong | 170 |
| The Chæropus | 55 | The Colugo | 175 |
| The Myrmecobius | 57 | The American Bison | 177 |
| The Marsupial Mole | 60 | The Yak | 181 |
| The Ornithorhynchus | 61 | The Cape Buffalo | 182 |
| The Peacock Pheasant | 69 | The Musk Ox | 183 |
| Lady Amherst's Pheasant | 70 | The Rocky Mountain Goat | 185 |
| The Ocellated Turkey | 73 | The Harnessed Antelope | 186 |
| The Red Bird of Paradise | 75 | The Prong-horned Antelope | 187 |
| Schrenck's Tanager | 77 | | |
| The Chauna | 78 | The Wapiti | 188 |
| The Condor | 79 | The Musk Deer | 189 |
| The Extinct Starling | 92 | The Chevrotain | 190 |
| The Bullfrog | 97 | The Collared Peccary | 196 |
| The Pipa | 105 | The Brazilian Tapir | 201 |
| The Amphiuma | 109 | | |

## LIST OF ILLUSTRATIONS

| | PAGE | | PAGE |
|---|---|---|---|
| The Common African Rhinoceros | 202 | The Narwhal | 324 |
| The True or Common Zebra | 206 | The Round-headed Porpoise | 329 |
| The Racoon | 212 | The Common Dolphin | 330 |
| The Coati-mundi | 214 | The Ring-tailed Lemur | 339 |
| The Kinkajou | 215 | The Short-tailed Indris | 340 |
| The Sloth Bear | 220 | The Long-tailed Indris | 341 |
| The Poiana | 231 | The Senegal Galago | 344 |
| The Arctogale | 233 | The Slender Loris | 345 |
| The Binturong | 234 | The Potto | 347 |
| The Foussa | 236 | The Tarsier | 348 |
| The American Badger | 239 | The Aye-aye | 349 |
| The Skunk | 241 | The Prairie Dog | 353 |
| The Two-toed Sloth | 252 | The Mole-rat | 357 |
| The Great Ant-eater | 254 | The Jerboa | 358 |
| The Smallest Ant-eater | 256 | The Tri-coloured Tree-Porcupine | 359 |
| The Apar Armadillo | 258 | The Russian Desman | 362 |
| The Long-tailed Pangolin | 260 | The Gymnura | 363 |
| The Aard Vark | 263 | The Dwarf Tupaia | 364 |
| The Californian Sea-Lion | 277 | The Typical Jumping Shrew | 365 |
| The Greenland Seal | 287 | | |
| The Sea-Elephant | 292 | The Potomogale | 366 |
| The Walrus | 294 | The Tailless Tanrec | 366 |
| The Dugong | 304 | The Solenodon of Cuba | 367 |
| The Manatee | 306 | The Golden Mole | 368 |
| The Greenland Whale | 315 | | |

# TYPES OF ANIMAL LIFE

I

## MONKEYS

IF any one knew when it was that the first flint-implement was struck out with a will by Palæolithic men, he might be able to tell us how long the period is since the monkey's resemblance to ourselves was first a subject of remark. Of that period, the time which has elapsed since the very old line, "*Simia quam similis turpissima bestia nobis,*" was written, can, at any rate, be but a small fraction. Of late the progress of knowledge has largely increased the interest, felt from of old, in this most exceptional group of animals. The more we know of science the more we know of their bodily resemblance to us and of their divergence from every other creature, and the more also do we become interested in their ways and in that problem which concerns their origin.

Most readers are probably, by this time, not a little tired of Darwinian controversy; and certainly we have no intention of dealing with it here. That would lead us into psychology and metaphysics, while our present purpose is to deal only with what appear to us to be the most interesting facts which concern their natural history, and not to analyse the phenomena of the ape mind.

Readers who may be interested in that very important question are referred to the author's recent work, entitled "The Origin of Human Reason" (Kegan Paul, Trench, Trübner & Co.). It is pretty certain, however, that were apes as like us mentally as they are bodily, that very similarity would result in a notable difference. Some men are Radicals and some Conservatives, but apes would give a solid vote for the most Conservative ticket, since that progress and advance of civilisation which pleases most of us means, ultimately, death to them.

Progress has indeed its drawbacks, even for the zoölogist and for every passionate lover of Nature. Since the days when Banks and Solander were carried by Captain Cook round the globe to explore new regions, what havoc has not been committed! That fair, new world upon which they gazed with admiration and wonder, such as we might feel could we visit another planet, is being rapidly deprived of its interesting animal inhabitants; even more, perhaps, through the pernicious agency of enthusiasts for "acclimatisation" than by the spread of agriculture or the multiplication of flocks and herds. The plains of Africa, which only half a century ago teemed with wild animals, are becoming a zoölogical desert, and the "common" zebra is now almost extinct, while the bison (so often called buffalo) would very soon be exterminated but for the protection of the autocratic empire of the East and the great republic of the West.

In the forests and jungles, the wide wastes and rocky fastnesses of the tropics, however, the agile ape will yet long hold his own on both sides of the Atlantic.

Every one knows that there are monkeys in tropical America, no less than in Africa and Asia; but few persons who are not naturalists know how strangely different are the species which inhabit the Old World from those of the New. The whole mass of apes of all kinds is, for the purposes of study, grouped in two families, each of which is considered as made up of smaller groups termed "sub-families"; and these again of "genera," each genus containing certain different kinds or species of apes. Now, no single kind of ape which exists in America is found anywhere else; so that all the above-mentioned groups are similarly confined either to the east or to the west of the Atlantic Ocean.

The Old World has given rise to the chimpanzee, gorilla, and orang, the long-armed apes, many long-tailed apes, and every species of baboon.

In the New World are found spider monkeys and howling monkeys, sapajous and sakis, the gentle night ape (dourocouli), the graceful squirrel monkeys, and those charming pigmies of the monkey world, the little marmosets. We have thus two great families of monkeys, one including all the above from the chimpanzee to the baboons; the other comprising the remaining forms, namely, from the spider monkeys to the marmosets.

These two families of Old and New World apes differ literally from head to tail. In most of those points in which they differ, it is the Old World forms which are the more like man; but, nevertheless, in some respects the Americans have progressed further than the denizens of Africa and Asia. They have developed an additional wisdom tooth, and no others can make so wise a use of their tails. There seems to be no forest region in the world comparable with that of Brazil; for the dreary one of Africa, described by Stanley, appears far inferior in the development of its trees. But in Brazil, as Alfred Wallace has so graphically described, forest is fitted to and superimposed on forest. At a great height a waving sea of verdure, rich with animal life, is spread out in the dazzling sunshine, borne up on columns which tower through the obscurity of the vast space beneath, wherein a second growth of what would elsewhere seem noble trees finds a congenial home. Beneath these, again, there may yet be another similar but smaller growth, while lycopods and a multitude of humbler herbs clothe the soil. Evidently, if adaptation to surrounding conditions takes place anywhere among animals, special adaptations to forest life may be looked for here; and here they are found. Many a bird and beast which elsewhere exists in plains or in woods of relatively small extent, has here its emphatically arboreal representative, as the fowl seems to be represented by the curassow and the goose by the horned screamer. For animals which cannot fly, but have to pass their lives amid such an ocean of forest, it is especially needful that they should be supplied with all possible means of avoiding a fatal fall.

Thus the sloth, which passes its life hanging beneath the branches, has its hands and feet changed into what seem mere hooks, which remain bent over when at rest and need an effort to unclasp. By this means the animal can sleep securely while hanging, back downward, within its leafy bower. Monkeys, as we all know, have the feet modified into prehensile organs acting like hands, the great toe grasping powerfully in opposition to the other four. This modification wonderfully adapts them for tree life. There is, however, one further possible adaptation, and only one, and it is just that very adaptation which is to be found in the monkeys of American forests. It is an adaptation which supplies them with what is practically a fifth hand. In the spider monkeys, the woolly monkeys and the howling monkeys, the under-surface of the terminal portion of the tail is naked, so that it can be very closely applied to any surface with which it is in contact. The tail itself is a very powerful organ, and is capable of curling its own end so firmly round an object that the animal's whole body can thus be safely suspended. A tail of this kind is called a "prehensile tail." Not every American monkey has it, but no monkey which is not American possesses anything of the kind. Its possession must greatly add to the security and ease of locomotion of any forest-dwelling beast. An amusing illustration of the widespread ignorance which exists as to such matters, and also of the use of the imagination in a way not strictly scientific, occurred with reference to the Prince of Wales's visit to India some years ago. Among other places of interest the Prince visited was the Temple of Monkeys at Benares. His visit was duly depicted in one of the illustrated journals, and no doubt with scrupulous fidelity in all those points to which the artist directed his attention. Nevertheless, these monkeys are represented as having prehensile tails; which is about as accurate as would be a picture of a fox-hunt by a supposed eye-witness, wherein the hounds should be represented each with a fox's brush for tail, Reynard himself bearing the curly caudal appendage of a thoroughbred pug.

But it is not only in form and structure that American monkeys are distinguished, but also in quickness of intelligence and gentleness of disposition.

At least many of those large animals, the spider monkeys, are singularly gentle, and such is especially the case with the little squirrel monkeys—perhaps the most attractive of the whole order to which they belong. As to intelligence, it is the commoner monkeys of South America, the sapajous, whereof itinerant Italians love to make use for tricks and performances. We can vouch from our own repeated observations, for the amazing skill and rapidity with which they will catch and return a ball, sweep their stand, load and fire their miniature gun, and play the various antics to which they have been trained.

Nevertheless, the monkeys of the Old World are, as we have said, the more man-like in structure, and no animal to be found between the Atlantic and the Pacific, makes any approach to the closeness wherewith the anthropoid (or specially man-like apes) resemble us.

Some Old World monkeys have no thumbs, and none have what we should call a good one; but even the most brutal baboon has a better one than has any of the American apes, in all of which the thumb is more like a fifth finger, bending around nearly in the same plane as the others.

Only one monkey has a chin, and that is an inhabitant of Sumatra, and no ape out of Asia has a really prominent nose.

No monkey's tones are so pleasant as are the flute-like notes which the sapajous will often emit when pleased, but no ape gives out such man-like sounds as are chanted by the long-armed apes, or gibbons, of the Old World. Often he has a short tail, and sometimes none, but almost all those of America have a long one, though in a few very singular forms, to be presently noted, it is short.

Monkeys of the Old World ascend to higher latitudes than do those of the New. None are known to us as having been found in America to the north of southern Mexico; but monkeys are denizens of Gibraltar, Central Asia, and Japan, in the eastern hemisphere. This great distinctness between the apes of the Old World and the New, at once suggests some curious questions to which, as yet, it is quite out of our power to return answers. Which group is the older? Did monkeys, as is commonly supposed, first exist in the Old World? If so, then did the American monkeys originate in the New World? and if not, how did they find their way into it and whence did they come?

It is a common opinion that the aboriginal tribes of American men are of Mongol affinity and migrated from Asia; but as this is a question we have not studied, we preserve an open mind concerning it. Although wild American dogs have been domesticated, Asiatic emigrants may none the less have brought domestic dogs with them. But however this may be, it is pretty certain they brought with them no monkeys. No fossil remains, so far as we know, at all justify a belief in the Old World origin of New World forms; so that up to the present time the relative age of the two groups and the origin of either of them are mysteries. Speculative opinions have indeed been formed favouring the notion that the American apes are specially related to certain creatures called Lemurs (to be seen in many zoölogical gardens) which have their headquarters in Madagascar. The evidence, however, upon which some naturalists rely as justifying this opinion, is, in our eyes, untrustworthy. Neither have we been able to detect any sign of the former existence of creatures of the monkey kind that were intermediate between those of the Old World and those of the New; so that it seems to us not improbable that the two groups may have had an entirely different origin, and that the points of structure in which they so remarkably agree, are but analogical resemblances and not signs of any special blood relationship between the two. Bearing in mind the great distinctness of these two families, we may now proceed briefly to review the leading forms contained in each, commencing with the Old World family, and especially with the species belonging to it which are most like ourselves—the anthropoid apes.

First of these, in popular estimation, stands the famed gorilla. Our knowledge of this largest of apes, knowledge both ordinary and scientific, is due to Americans. It was really discovered by Dr. Thomas Savage, who, with the assistance of a missionary named Wilson, procured materials sufficient to enable Prof. Jeffries Wyman to describe important parts of its anatomy. (See the *Boston Journal of Natural History*, vol. iv. 1843-4, and vol. v. 1847.) That absurd dogma which has been defined and decreed by leading agnostics, the dogma that "every man receives microscopic justice in this world," can be pretty well refuted by the history of physical science. In geography we have one notable instance. Christopher Columbus, with a hardihood now difficult to realise, sailed across an utterly unknown ocean and discovered a new Continent, which, nevertheless, has not been named after him, but after his imitator, Amerigo Vespucci. In another branch of science we often hear something about galvanism, and sometimes use the term. That curious kind of force received its name from Galvani, who called attention to it in 1789; but Swammerdam had discovered it one hundred and thirty years earlier.

The last biological novelty is the hypothesis that every organism, however long or short its life may be, contains an immortal substance, transmitted from generation to generation, and from century to century. Every one now couples with this idea the name of Prof. Weismann, ignoring the fact that the same doctrine was publicly taught by poor old Sir Richard Owen half a century ago.

Similarly, the discovery of the gorilla has been so generally attributed to Mr. du Chaillu, that justice makes it needful to remind our readers of the debt due to American discoverers and describers of a preceding generation.

Nevertheless it seems probable that the animal had been described, and even specimens of it obtained, more than two thousand three hundred years earlier. In 510 B.C., the Carthaginian Government decreed that thirty thousand persons should be transported south of the Pillars of Hercules to found Phœnician colonies on the West African coast. Hanno set out accordingly with a fleet of sixty vessels, and subsequently read a report of his expedition to the Senate at Carthage.* Therein he stated that after a long journey they entered a gulf, and near it found a number of "wild men" entirely covered with hair, who were called "gorillas" by his interpreters. "We pursued them," he says, "but could not take any of the men on account of their quickness in climbing, but we took three women, who bit and tore those who carried them off, so that we were obliged to kill them. We then skinned them and carried their skins home with us." Two of these, stuffed, were placed by Hanno in the temple of Astarté at Carthage, where they remained till that city was captured by the Romans.

The extent of Africa inhabited by this animal is not large, only including the forest region, extending inland between the mouths of the Cameroon and Congo Rivers

Its smaller cousin, the chimpanzee, is found over a much wider range, namely, from the Gambia to the Benguela, extending inland as far as 28° East longitude. The earliest notice of what was probably this animal

* "Ann. des Sc. Nat." (third series), Zoology, vol xvi., 1851, pp. 184-188.

appeared in a description of the kingdom of Congo by Pigafetta in 1598, published at Frankfort. A further notice appeared in a curious book, entitled "Purchas: His Pilgrimage," in 1613; but in the last year of the same seventeenth century a full and accurate account of the structure of the chimpanzee, with excellent plates, was published in London by Tyson under the title "Anatomie of a Pigmie."

The gorilla has been hardly seen in Europe, though a specimen lived for a time in the Westminster Aquarium, and in Berlin; but the much smaller chimpanzee has often been exhibited alive in London, and is an attractive feature in menageries, not only from its resemblance to a child deformed by preternatural wrinkles of age, but also from its liveliness and the facility with which it acquires a number of playful tricks.

There was till lately a chimpanzee, known as "Sally," at the London Zoölogical Gardens which was in three ways remarkable. To begin with, it was the best grown and largest specimen which has lived in Europe; secondly, it differed from all before known ones by its carnivorous habits. It would greedily seize and devour small birds, whereas such apes were previously supposed to be naturally vegetarians only. But it was, in the third place, most remarkable for the tricks it acquired. It would separately pick up from the ground, place in its mouth, and then present in one bunch, two, three, four, or five straws, as might be demanded of it, or only one. It had distinctly associated the several sounds of these numbers with corresponding groups of picked-up straws. It would also, on command, pass a straw through a large or small hole in the fastening of its cage or through a particular interspace of its wire netting. Finally, it would, when so bidden, put objects into its keeper's pocket, play various odd tricks with boy visitors, and howl horribly when told to sing.

A great contrast to the African chimpanzee is presented by the third anthropoid form of ape, the Asiatic orang. It is red instead of blackish in colour, and its arms are so long that they reach to the ankle when the animal stands erect. This it rarely, if ever, does spontaneously. It walks resting on its knuckles and the outer edges of its feet, their soles being turned inward. Thus resting on its hands, it uses its arms as a pair of crutches, swinging the body and legs forward between them. Its disposition is also very different from that of the lively and petulant chimpanzee. Remarkably calm, not to say languid, in its actions, it has in captivity a curiously melancholy demeanour. Its high rounded forehead, very different from the villanously low brows of the chimpanzee and the gorilla, gives it a singularly intellectual aspect, so that when we observe it pensively squatting, with fat belly—like an image of Gautama—we might fancy that the mind of some esoteric Buddhist was imprisoned within the apish body, incapable of making its latent existence known, and mutely contemplating the longed-for Nirvana.

The orang is found nowhere in the world except in Sumatra and Borneo, and even there only in lowland humid forests, which supply it at once with shelter and the vegetable food it loves. A solitary and peaceful animal, it is ordinarily very slow and deliberate in its movements. Nevertheless, when attacked, it can defend itself with alacrity and effect, as the following anecdote (from Wallace's "Malay Archipelago") will show : "A few miles down the river there is a Dyak house, and the inhabitants saw a large orang feeding on the young shoots of a palm by the river-side. On being observed, he retreated towards the jungle, which was close by, and a number of the men, armed with spears and choppers, ran out to intercept him. The man who was in front tried to run his spear through the animal's body, but the orang seized it in his hands, and in an instant got hold of the man's arm, which he seized in his mouth, making his teeth meet in the flesh above the elbow, which he tore and lacerated in a dreadful manner. Had not the others been close behind, the man would have been more seriously injured, if not killed, as he was quite powerless: but they soon destroyed the creature with their spears and choppers. The man remained ill for a long time, and never fully recovered the use of his arm."

The only other specially man-like, or anthropoid, apes are the long-armed apes or gibbons. They are generally much less thought of by the public than those more celebrated creatures, the gorilla, chimpanzeee, and orang. Nevertheless, they present several points of great interest, and in some respects more resemble ourselves than does any one of the three kinds just mentioned. The gibbons are smaller creatures, but are all as completely destitute of a tail as are their three more renowned relatives. The largest gibbon stands about three feet high from head to heel. There are several species, but they vary so much in colour, according to age, sex, and other conditions, that they cannot yet be said to be very well-defined. They range over south-eastern Asia, and at present are nowhere else found; but in Tertiary times a gibbon, much larger than any now existing, roamed through the forests of the south of France. Though some are to be found in India, and others in Burmah, Malacca, and Siam, their special abode is the Indian Archipelago, in the great islands of Borneo, Sumatra, Java, and in others, for they are there widely diffused, save in the Philippine Islands. Their agility is most wonderful. They will swing themselves from

FIG. 1

THE SIAMANG GIBBON.

branch to branch by their long arms, with such amazing rapidity that they seem almost to fly through the forest. We have often watched their wonderful motions in a large cage specially provided at the Zoölogical Gardens. We have also listened to the remarkable manlike sounds they emit (as before said) when singing or shouting, as they so often do. In spite of their great activity these animals are exceedingly gentle and make excellent pets, although they can inflict terrible wounds with their elongated canine teeth. The siamang (Fig. 1), which is the largest of the gibbons, inhabits Sumatra, and goes about in troops there, making the woods re-echo morning and evening with its deafeningly sonorous cries.

The gibbons have arms so long that they reach the ground while the body is perfectly erect. This extreme elongation of the arms tends to prevent our noticing the really great proportional length of their legs. In nothing do the gorilla, chimpanzee, and orang differ from man in structure more notably than in the shortness of their lower limbs. The gibbons go to an extreme the other way, for if the leg be compared with the body as to its length, then the gibbons have even slightly longer legs than man himself. This is a very noteworthy approximation to human structure. There is yet another. We have already said that only one monkey, a gibbon, has a chin. That monkey is the just mentioned siamang. Its chin is more developed than is that of not a few human beings. In spite of these approximations they have one noteworthy falling off behind. The body is bare where it rests on the ground in a sitting posture, and the hardened naked patches of skin thus situated are spoken of as ischiatic callosities on account of the bones (ischia) which they invest. In possessing these callosities they agree with all the other monkeys of the

Old World, save the gorilla, chimpanzee, and orang, which have them not. Neither have any American apes, which, more decorous than their Transatlantic brethren, have that region of the body copiously clothed with hair.

Leaving now the anthropoid apes, with which alone we have been hitherto occupied, we find when we pass to the next group of monkeys a remarkably different aspect and a very different form of body. The limbs are nearly of equal length, but the arms are now the shorter ones, so that a quadrupedal mode of progression on the ground is natural to them. Nevertheless, they are arboreal animals, and both adroit climbers and dexterous jumpers. They are aided in keeping their balance during their movements by the possession of a long tail.

The first group of these monkeys is one which is found only in central and south-eastern Asia, and consists of many species which also have their headquarters in the Indian Archipelago. They are not very often seen in captivity save that well-known kind, the entellus monkey or hounaman, which is an object of such religious veneration on the part of the Hindoos. It has a coat of whitish hair but a jet-black face, and once seen is not likely to be forgotten. The largest and by far the most singular species of the group, however, has never been seen alive in Europe. It is exclusively a native of Borneo, where it can hardly be common, since, though it was figured and described by Buffon in 1789, it has found its way to no menagerie. This very remarkable beast is the kahau or proboscis monkey (Fig. 2), which differs from every other ape in having a long projecting nose. Two fine stuffed specimens of this creature are to be seen in the British Museum, one young, the other adult. The young of this species instead of having a nose similar to that of the adult save as regards size, have it relatively much shorter, and also turned upward. A zealous

FIG. 2.

THE PROBOSCIS MONKEY.

and learned Lazarist missionary, a Frenchman, the Rev. Father David, has made many important zoölogical discoveries in central Asia. There, high up in the cold forests of Moupin, in Thibet, he found an ape clothed with dense fur, suitable for its frigid abode. It lives in a region where frost and snow last several months in the year, and where it has little to eat but the shoots and twigs of trees. Nevertheless, this ape, living in a region so remote from Borneo, with its hot, humid forests, is very like the young form of the proboscis monkey. It differs from the latter, however, in having a nose turned up to the highest possible degree, on which account its describer, Prof. Alphonse Milne Edwards, named it "the monkey of Roxallana," in honour of that "tip-tilted" imperial beauty.

The Indian monkeys, which in general structure are most like the kahau and the entellus, are closely resembled by the species of an African group, the members of which are called colobi. These African apes have had a too fatal popularity, the glossy coats of their well-clothed skins having been for a time the favourite material for ladies' muffs, the well-known "monkey muffs." Several species of colobi are very notable for their wonderfully handsome fur fringes or tippets of long white hair, accompanying a general livery of the deepest black. Their Indian allies have very feebly developed thumbs, but the colobi are remarkable for having no thumbs at all. A specimen presenting this condition of hand may well seem to a non-scientific observer as one accidentally or purposely mutilated. We recollect a few years ago having our attention arrested by two very fine specimens of this genus which we saw mounted in a tobacconist's window in London. To our surprise we observed that they had thumbs, and so we at once entered the shop, and asked to be allowed to inspect them. We then found that artificial thumbs had been

B

sewed on, and the proprietor of the shop admitted that he had had this done to restore the specimens to what he supposed must have been their natural condition!

It is often assumed that wild animals escape most of the evils to which civilised human flesh is heir. No doubt in most cases when such creatures are sufferers from disease, merciful Nature calls in her destructive powers to make a speedy end of their sufferings. Nevertheless, skeletons in our museums show that these apes, in their hot, damp native forests, do occasionally suffer severely from acute rheumatism.

The two sets of long-tailed apes just noticed, form together one very natural and distinct section or sub-family of the ape order. With the exception of the lofty region of Thibet, they are confined to the warmer parts of Africa and Asia, although in Miocene times they ranged through Europe from Greece to Montpellier, if not farther north.

The next sub-family of monkeys to be here noticed is one which is no less distinct, though the forms it contains are more varied. In it we find, as it were, a sliding scale of forms descending from graceful little African monkeys, such as the Diana monkey, the Mona or the white-nosed monkey, to the largest and most brutal of the baboons. The whole sub-family is imperfectly divisible into three groups. The first of these is made up of species exclusively African, such as the three kinds above mentioned, and their allies. The Diana monkey is so called from its white concentric band of hair above the forehead. The Mona is remarkable for its brilliant coloration, its head being yellowish olive with a black stripe on the forehead, yellow whiskers, and a purple face. The back is chestnut brown, and there is a white

MONKEYS 19

spot on each side of the root of the tail, which is black. Various species of this group are distinguished by curious

FIG. 3.

THE WHITE-NOSED MONKEY.

markings on the face. The white-nosed monkeys are very attractive, generally gentle animals, and most easy to distinguish by the character their name denotes. The moustache monkeys have a hardly less conspicuous stripe where the moustache should be. Other monkeys of western Africa are singularly distinguished by having the eyelids white, though the rest of the face is dark, and they are named accordingly " white-eyelid monkeys." The commonest of the whole group is the green monkey, which naturally inhabits the Cape Verd Islands, but which has been introduced and has run wild in one of the Antilles.

All these African monkeys have long tails and ischiatic callosities. They also have better developed thumbs than the Asiatic ones of the group containing the entellus, and they introduce us to a new character. If a visitor to a menagerie presents one of these small African monkeys with first one and then another nut, the nuts will not at once be cracked and eaten, but will be put successively inside the cheeks, which will be observed to protrude in a remarkable manner. These dilatable face pockets are called cheek-pouches. Nothing of the kind is possessed by any of the higher Old World apes, though their possession is a constant character not only of the group we are now considering, but also of all the Old World monkeys which yet remain for us to notice, and which may be taken to constitute two more groups. The first of them is entirely confined to the continent of Asia, with the one exception of the Barbary ape, which also lives on the rock of Gibraltar. The existing specimens there abiding are, however, either individuals which have been of late reintroduced from Africa, or they are the offspring of such. The Barbary ape, or "magot," has a special interest, from the fact that a time when prejudice did not allow the human body to be used for medical study and dissection, the body of that ape was employed as a substitute, as very old anatomical works conclusively prove. The remaining, and exclusively Asiatic, members of the group are known as "macaques." Some of them have long

FIG. 4.

THE WANDEROO.

tails, some short tails, and a few, like the Barbary ape, have none. They extend farther north than other monkeys, namely, to Japan and northern China, and one species was found by Father David at Moupin in Thibet. Including the Barbary apes, this group may be said to be most widely spread of all those which compose the monkey order, extending in one direction from Gibraltar and northern China, down to the island of Timor and

FIG. 5.

THE BLACK MACAQUE.

the Cape of Good Hope, and in another direction from north-western Africa to Batavia, Japan, and the Philippine Islands in the east. In ancient times this genus extended into France, and even to England. One Indian species, called the wanderoo (Fig. 4), has the face encircled by very long hairs, which gives the ape a very conspicuous and characteristic appearance. It is also somewhat distinguished as regards its domestic habits, if there be truth in the following singular reproach cast upon the Veddahs of India by some of their fellow-countrymen: "The Veddahs are like wanderoos: they have only one wife!"

FIG. 5*a*.

THE CHACMA.

In the island of Celebes an exceptional kind of ape is found called "the black macaque," which by its structure leads us on to the lowest group of Old World species, the baboons, in spite of the remoteness of the region it inhabits from that which is their home. The baboons are exclusively inhabitants of Africa and of Arabia — which is considered, as regards its animal population, to really form a part of Africa. The baboons are the largest apes, after the anthropoid ones. They are also the most quadrupedal in their mode of propulsion, and have by far the most prominent muzzle, being known as the *cynocephali*, or dog-headed apes. The ape with the most exaggerated snout is the chacma (Fig. 5a) of South Africa. It is a very powerful brute, which lives in troops among rocks, and, though mainly a vegetable feeder, will also eat eggs, large insects, and scorpions, which it is said to deprive of their sting by a very sudden and dexterous pinch.

One of the most singular of the baboons is the mandrill, which exceeds the chimpanzee in bulk of body. It is remarkable for the brilliant coloration of its face, the cheeks being brilliant blue, the nose vermilion, and the beard golden yellow. It was an example of this species which, in the earlier years of this century, was known at Exeter Change as "Happy Jerry," and used to smoke his pipe and drink his glass of gin and water before admiring visitors. The venerable relics of this felicitous ape are still carefully preserved in the national collection at the British Museum.

With the baboons, we end the series of Old World forms. In America we can find nothing which at all closely resembles them. It has been suggested that the howling monkeys, the largest in bulk, may be taken as the representatives in the New World of the baboons of the Old, but we know nothing, either in their organisation or their habits, which really warrants the suggestion. Instead of being rock dwellers and quadrupedal in their locomotion, howling monkeys are extremely arboreal, and are one of those groups which, as before observed, are furnished with a perfectly prehensile tail. As their name implies, they are most noted for their prodigious

# MONKEYS

cries, which are said to be sometimes almost deafening. A curious modification of structure goes with this por-

Fig. 6.

THE LONG-HAIRED SPIDER MONKEY.

tentous clamour. At the root of the tongue in ourselves, and in beasts generally, is a small, solid, transversely extended bone. In these apes the bone in question is of enormous size, as it were swollen into a great bony bladder with very thin walls. There can be little doubt but that the resonance of their voice is enormously augmented by this bony drum. In captivity, howling monkeys seem sullen and morose, and, though not petulant, have none of that gentle amiability which is to be found among the next group of American apes, the spider monkeys.

These latter (Fig. 6) are no less wonderfully adapted for tree life, while they are more active, and seem to represent to a certain extent in the New World, the long-armed apes of the Old, although they are very slow animals compared with the gibbons. Long arms they have indeed, and also legs, whence their name; but the former do not predominate over the latter at all, as in the gibbons. So powerful is the grip of the spider monkey's prehensile tail, and so dexterously is it used, that not only can the animal's whole body be sustained by means of it, but it even serves as a fifth hand, grasping and bringing to the mouth or paws objects otherwise out of reach. Their prehension in some other respects is singularly defective, as they alone among American monkeys resemble the colobi of Africa, in having no thumbs, or only a minute rudiment of one. They have no cheek pouches, nor has any other New World ape, and no one of them (as has been already mentioned) has ischiatic callosities.

Certain monkeys known as woolly monkeys closely resemble those just described, save that they have well-developed thumbs.

Next comes the group composed of those commonest and most frequently seen of the New World apes, the sapajous, already referred to as being so much in request for tricks and exhibitions. They are considerably smaller in size than the spider or howling monkeys, and make good pets, grinning with the most curious grimaces and uttering flute-like sounds when responding to caresses or endearments. They are very numerous, and there are probably at least some twenty different species, though they vary so remarkable in colour that their real number is by no means satisfactorily determined. It is possible that in the sapajous and in the howling monkeys we have groups of animals wherein new species are now in actual process of formation. That careful naturalist, Dr. Rengger, managed to obtain some rare opportunities of observing these watchful, timid animals in Paraguay, which is about their southern boundary. He tells us that they spend their lives almost constantly in trees, which they only quit to drink at some spring or stream, or to forage in some tempting field of maize. Sleeping at night between conveniently branching twigs, they pass the day ranging from tree to tree in search of fruit, buds, insects, honey, or for some nest's eggs or callow brood, going about in family groups of from five to ten individuals. On one occasion a large troop approached him while he was hidden from their observation. First came an old male, passing from one tree's summit to another, and keeping a careful look-out in all directions. He was followed by about a dozen others of both sexes, three of the females each carrying a young one on the back or under the arm. One monkey quickly espying a neighbouring orange-tree covered with ripe fruit, with a loud cry sprang upon it, followed by all the others, who immediately fell to work, some remaining on it to enjoy their feast, while others retreated to adjoining trees, there to enjoy in quiet the booty they had secured. Sapajous are to be found in every menagerie, and if one happens to be observed in proximity to one of the Asiatic or

African monkeys, the characteristic difference between the countenance of Old and New World species will be at once apparent. In those of the Old World the nose is not only (save in the proboscis monkeys) very small, but also very narrow, its nostrils being in close proximity. In the sapajou and the New World forms, however, the nose is remarkably broad, the two nostrils being widely separated. In leaving the sapajous we bid adieu to the groups furnished with prehensile tails, and come upon a very different set of forms, which terminate the order of apes.

First we may notice the group of monkeys of about the size of sapajous, and known as sakis. They are somewhat widely spread over the South American continent, but are nowhere very abundant, living in pairs, alone, or accompanied by their young. They are gentle, timid animals, which sleep much by day and go abroad at night, thus escaping the persecution they would otherwise suffer from the oppression of the more active and powerful sapajous. Sakis are seldom seen alive in captivity, but several of them are very singular in appearance. One is known as the "Capuchin," on account of its brown colour and long beard; while another kind, also provided with a beard, has, on account of its fine black colour, been called the Satanic saki. Another species is black, with a white head, while another has its pate more or less bald (Fig. 7). This last-mentioned kind agrees with some others in having a very short tail. In this they differ not only from the other sakis, but from all the rest of the monkeys of America, every one of which is provided with an elongated caudal appendage. One of these exceptional apes has the tail not only extremely short, but furnished with long hair, so that it forms a prominent ball which would serve as an excellent "dress improver," were there only a dress to improve. It inhabits the upper part of the enormous valley of the Amazon. The young traveller, Delville, whose premature death was a sad loss to science, tells us

FIG. 7.

THE BALD-HEADED SAKI.

that a specimen of this species which he captured had the frequent habit of rising spontaneously and walking erect, and that it soon learnt to drink from a glass held in its hand, drinking regularly twice a day. It was very fond of milk, bananas, and sweetmeats, but had, unlike "Happy Jerry," a horror of tobacco, snatching a cigar from the mouth of any one who sent smoke towards it and grinding it on the ground. When several bananas were given it, it held one with its hands and the rest with its feet. It was gentle and affectionate to its master and some others, and liked to lick their hands or face; but it was very hostile to a young Indian, and when in a passion would rub its hands together with extreme rapidity.

The douroucoulis, or night apes, are, as their name implies, truly nocturnal animals, passing the whole day rolled up asleep, within some hollow tree. Their great eyes, which are said to be luminous at night, seem to suffer much from strong daylight. Humboldt, who kept one for five months, tells us that it slept regularly for from between some time after dawn (at nine o'clock at the latest) till seven in the evening. At night they are as active as other apes are by day, and will make a great noise with their cries. They are reported to be exemplary monogamists.

We have more than once spoken of those graceful little animals, the squirrel monkeys or "saimiris." They are slender in form, with pretty rounded faces and long heads, which contain more brains in proportion to their bulk than does the skull of man himself. Their brilliant colouring also makes them attractive, and they are said to be affectionate and sensitive as well as gentle, their eyes filling with tears if treated harshly. They are greedy pursuers of insects, and have a somewhat singular taste, as spiders, which they are very dexterous in catching, are their supreme delight.

The last set of monkeys we have to enumerate is one which differs greatly from all those hitherto noticed. It is composed of the marmosets, or ouistitis, a numerous group of very small animals exclusively confined to the warmer parts of America. All the monkeys yet described by us have nails, which, though more convex and pointed than those of man, are nevertheless substantially like his. The marmosets, however, have all their fingers and toes provided with long pointed

FIG. 8.

THE SQUIRREL MONKEY.

claws, with the single exception of the nail of each great toe, which is flattened like our own. All the monkeys hitherto noticed, except the colobi and spider monkeys, which have no thumb at all, have a thumb which is more or less opposable to the fingers, although very imperfectly so in the American forms. The thumbs of the marmosets, however, are not opposable at all. The monkeys hitherto noticed, whether they have two or three wisdom teeth on either side of each jaw, all agree with us in having three grinders, which have milk predecessors, and are technically known as premolars. But the marmosets alone have only two such on either side of either jaw.

Various species of this pigmy group, each about the size of a squirrel, or even smaller, are remarkable for the beauty of their furry coat. Thus the "marikina," or silky marmoset, is clothed with fur of a golden yellow, that of the head and shoulders being long and forming a sort of mane. The pinche has the hair of the head greatly elongated. Another kind has a dark body, while the hands and feet are clothed with bright red hair. Several species have a long tuft of hair projecting outward from either side of the head.

We have compared these animals with squirrels as regards their size, but they may be similarly compared with respect to their movements among trees, to which they cling with their sharp-pointed claws, just as squirrels do. But though they thus resemble squirrels as to their mode of motion, their activity is by no means so great. They live in small troops, feeding on fruit and insects, which, like the saimiris, they eat greedily. They are very delicate in constitution, so that when brought into northern climes they generally live but a short time. Nevertheless, they occasionally breed in Europe, and bring forth as many as three at a birth, while all the other apes habitually bring forth but one. The father shares, at least sometimes, the mother's parental cares with great amiability, taking the young ones from her and carrying them about, when she is fatigued, till they need another supply of food.

With this notice of the marmosets ends our short

review of the entire group of apes, an order of animals consisting, as we have said, of two sections, each made up of the various subordinate groups which have now been enumerated and briefly noticed. Considered as one whole, the ape order ranges through the warmer part of the earth, from Gibraltar and northern China, to the Cape of Good Hope and the Island of Timor, in the Old World, and from 23° North latitude to about 30° South latitude in the New World. Individuals of the entellus monkey group have been seen near Simla at an altitude of 11,000 feet. Some of the localities richest in monkeys are islands, as Ceylon, Borneo, Sumatra, Java, Fernando Po, and Trinidad.

There are, however, certain islands which seem eminently well suited to support a large ape population, where apes are, nevertheless, conspicuous by their absence. Such are Madagascar, New Guinea, and the West Indies. Moreover, no ape is found even in the most tropical or best wooded parts of Australia. It is the more remarkable that no ape should be found in the great island of Madagascar, so rich in forests, seeing that it is the special home of those beasts, before referred to, named lemurs, which have generally been supposed to be very closely related to the apes.

In one or two of the West Indian islands monkeys have been introduced and have run wild, showing that they could very well have lived there had they been able to enter the Antilles without the aid of man. Trinidad is not a West Indian island. It is a detached portion of the South American continent.

We come now to the question concerning the existence of any special resemblance between these animals and some other given order of beasts. This is a problem by no means easy to answer. As we have said, apes are commonly thought to be most nearly approached by the lemurs, but the advance of anatomical knowledge hardly favours that view. Such resemblance is mainly due to the formation of the extremities. Lemurs, with one exception, certainly have opposable thumbs and great toes to both hands and feet; but opossums have feet with opposable great toes, and yet no one supposes that there is even the faintest special affinity between an ape and an opossum. In brain structure and in the more intimate processes of reproduction (generally deemed a valuable test of affinity) the apes and lemurs stand far apart; and on the principles of evolution we are convinced that there can be no close relationship between them, although it has been hastily assumed that lemurs were the direct ancestors of apes. Apes, in the present day, stand as it were on a sort of zoölogical island, and we have no clear evidence indicating from what neighbouring strand they may be conceived to have entered upon it. Their origin thus still remains wrapped in mystery. Nor is it clear that the apes of the New World and those of the Old ever had any ape ancestors common to both. Possibly further discoveries in the Eocene deposits of North America (which are such veritable treasure-houses of relics of ancient life) will reveal to us the past existence of transitional forms between the monkeys of America and of Asia and Africa; but, in spite of all that has been published, this has not, to our minds, been done, and we think it quite possible that these two families have had different origins, and have come to resemble each other independently. The possibility of "the independent origin of similar structures" is a doctrine we maintained in the first work we ever published, and increased knowledge and experience has more and more convinced us we were right in maintaining it. But whether the monkeys now existing on both sides of the Atlantic have had a single or a bifold origin, there can be no question but that they together constitute one very distinct and natural group, a group which, on account of the obvious and unquestionably close resemblance to ourselves of the creatures which compose it, must be ranked in the highest order of animals which exists, or, so far as we know, has ever existed.

## II

## THE OPOSSUM

WHAT is an opossum? That is a question well worth answering.

Opossums are animals which should possess a special interest for Americans, seeing that though the bulk of the species inhabit quite another quarter of the globe, it was nevertheless America which first made known to us any one of them. The one which was then first made known has, however, claims on the attention of dwellers on both sides of the North Atlantic. This claim is based on the fact that in very ancient times, as the immortal Cuvier discovered, it was an inhabitant of Europe.

It was in Paris (at Montmartre) that were first found the relics of these ancient animals, which still have living representatives in America. To these still living animals and to their allies we now propose to direct the readers' attention by doing our best to present them with an answer to our initial question, " What is an Opossum " ?

Doubtless most readers have some notion of the animal, which is still to be found in the south-eastern parts of the United States, and is known as the Virginian opossum; nevertheless, a few descriptive words may not be out of place here.

It is about the size of a cat, with a pointed head and

flesh-coloured nose, and generally white head and more or less black ears. The tail is about as long as the body, clothed with fur for nearly a quarter of its length, the rest being naked and covered with scales like the tail of

FIG. 9.

THE VIRGINIAN OPOSSUM.

a rat. But, unlike the tail of a rat, it is prehensile, thus resembling the tails of various American monkeys.

The body is clothed with long, loose, rather soft hairs some white, some black, some parti-coloured. Beneath the hinder part of the body there is, in the female, a pouch, wherein are the nipples—an odd one in the middle, the rest forming a circle round it.

The soles of the paws are naked, and all the toes have claws except the inner one of the hind foot, which is clawless, and acts like a thumb. In the front of the jaws are ten small teeth above and eight below, and there are seven grinders on either side of each jaw, one more wisdom tooth than even any American ape has.

The opossum is very destructive to poultry, but it robs birds' nests and feeds on fruit, its tail and hind feet being as well suited for climbing trees as those of the spider monkey. The expression "playing 'possum" refers to its habit of feigning death when pursued and overtaken. We have been assured that it will on such occasions endure much pain before it will exhibit any signs of life. The female brings forth from twelve to sixteen young, making a nest of dry grass in some hollow tree, or at a convenient spot among a tree's boughs. The newborn young are said to weigh scarcely more than a grain. Though they are naked, blind, and little defined in shape, they manage to find the mother's teats, to which they attach themselves firmly.

In about five days they grow to the size of a mouse and become shapely. Soon they will quit the pouch for a time, returning if frightened, and also to feed. While thus sheltering her young the mother will endure any torture rather than allow her pouch to be opened.

After this preliminary notice of the creature called the opossum, we may proceed to consider what its nature is, and wherein its most interesting peculiarities, as revealed by modern science, really consist.

When America was first visited by the explorers who succeeded Columbus, the opossum was noted as among its novelties, by R. Hamor in his description of Virginia, and by Hernandez, whose history of Mexico was printed in 1626, and by many others. But not then or till more than 200 years afterward did even men of science appreciate how great a novelty had been met with in this new animal. The habit of carrying its young in a pouch, and the fact that the hind paw (as in monkeys) was formed like a hand, were duly noted. The anatomy of a female specimen was described by Tyson in 1698 under the title of the "Anatomie of the Oposum," the year before he published his "Anatomie of a Pigmie," referred to before. A male specimen was also described by William Cooper in 1704. But however fully the creature's structure was investigated, the significance of its organisation remained hidden from men's eyes.

The exceptional nature of those flying beasts, the bats, was appreciated early, but not that of those marine beasts, the whales and porpoises. At the coronation feast of King Richard II., which took place in Westminster Hall on a Friday, roast porpoise figured on the board among the other dishes to be eaten on a day of abstinence, when flesh meat was forbidden food. Little did the first observers of the opossum imagine that the difference between a bat and a mouse, or that between a porpoise and a sheep, was as nothing compared with the difference which existed between the opossum and the racoon, or between the opossum and any other beast then known in Europe or America.

In order that this should be understood it needed that other creatures more or less allied to the opossum should be elsewhere discovered. The first sign of the coming revelation was the mistaken opinion that the opossum existed in the East Indies. Pison, in his history of Brazil, says that such a breed was called "cous-cous;" and Seba, in the middle of the eighteenth century, stated that an animal of the kind had been sent him from Amboyna, where its name was "coes-coes." For this, Buffon, in the tenth volume (1763) of his "Natural History," takes Seba roundly to task, asserting that the animal may have been first sent to Amboyna from America.

Yet earlier, in 1711, a Dutch traveller, Cornelius de Bruins, saw a creature which we now know to have been a kind of kangaroo. This was at Batavia, where several kangaroos were kept in captivity. He gave a fairly good figure of it in that account of his travels through Muscovy, Persia, and India, which was published at Amsterdam in 1714. This publication seems, however, to have excited little attention.

Rather more than half a century after this work appeared, the Royal Society of England took a step which was the starting-point of those discoveries which have resulted in enabling us at last to understand what an opossum really is.

At the recommendation and request of the Royal Society, Captain (then Lieutenant) Cook set sail in May 1768, in the ship *Endeavour*, on a voyage of exploration, and for the observation of the transit of Venus of the year 1769, which transit the travellers observed from the Society Islands on June 3 of that year.

Thus it was that 120 years ago a kangaroo was first observed distinctly and unequivocally by Englishmen. For, in the spring of 1770, Cook's ship started from New Zealand for the eastern part of New Holland, visiting, among other places, a spot which, on account of the number of plants found there by Mr. (afterward Sir Joseph) Banks, received the name of Botany Bay. Subsequently, when detained in Endeavour River (about 15° South lat.), owing to the need of repairing a hole made in the vessel by a rock (part of which, fortunately, itself stuck in the hole it made), Captain Cook tells us that on Friday, June 22, of that year, "Some of the people were sent on the other side of the water, to shoot pigeons for the sick, who, on their return, reported that they had seen an animal, as large as a greyhound, of a slender make, a mouse colour, and extremely swift." With respect to the following day he tells us: "This day almost everybody had seen the animal which the pigeon shooters had brought an account of the day before; and one of the seamen, who had been rambling in the woods, told us on his return that he verily believed he had seen the devil. We naturally inquired in what form he had appeared, and his answer was, says John, 'As large as a one-gallon keg, and very like it. He had horns and wings, yet he crept so slowly through the grass that if I had not been afeared, I might have touched him.' This formidable apparition we afterward, however, discovered to have been a bat. Early the next day," Captain Cook continues, "as I was walking in the morning at a little distance from the ship, I saw myself one of the animals which had been described. It was of a light mouse colour, and in size and shape very much resembling a greyhound. It had a long tail, also, which it carried like a greyhound, and I should have taken it for a wild dog if, instead of running, it had not leapt like a hare or deer." Mr. Banks also had an imperfect view of this animal, and was of opinion that its species was hitherto unknown. The work contains an excellent figure of the creature. Again, on Sunday, July 8, being still in Endeavour River, Captain Cook tells us that some of the crew "set out with the first dawn in search of game, and in a walk of many miles they saw four animals of the same kind, two of which Mr. Banks's greyhound fairly chased, but they threw him out at a great distance by leaping over the long thick grass, which prevented his running. This animal was observed not to run upon four legs, but to bound or leap forward upon two, like the jerboa." Finally, on Saturday, July 14, "Mr. Gore, who went out with his gun, had the good fortune to kill one of these animals which had been so much the subject of our speculation;" adding, "This animal is called by the natives *kangaroo*. The next day (Sunday, July 15) our kangaroo was dressed for dinner, and proved most excellent meat." Such is the earliest notice of this animal's observation by Englishmen.

Thus were we introduced to the knowledge of a creature which at first could not be, and was not, expected to have any special affinity to an animal so unlike it externally, and an inhabitant of so distant a country, as is the Virginian opossum.

As, however, the knowledge of Australia increased, it soon became evident that it was inhabited by a great number of various kinds of beasts, every one of which was quite different from all other beasts previously known, with the exception of some bats, a rat, and the Australian dog, the dingo. But, before proceeding to review the peculiar animals then discovered, it will be well briefly to take stock of all the various kinds of beasts previously known. All kinds of beasts taken together are considered by naturalists to form one class, the class of "backboned animals which suckle their young," or the class of mammals—mammalia. Other classes of backboned animals are birds, reptiles, and fishes. Every class is divisible into certain great groups called "orders," and in this way the class of beasts is subdivided into a number of orders such as the following :—1. Apes. 2. Bats. 3. Insectivorous beasts (such as the mole,

## THE OPOSSUM

hedgehog, and shrew mouse). 4. Carnivorous beasts (such as the lion, dog, bear, otter). 5. Seals and sea-lions. 6. Gnawing beasts or rodents (such as porcupines, hares and rabbits, rats, squirrels, jerboas, &c.). 7. Hoofed beasts (such as horses, swine, deer, oxen, antelopes, goats, sheep, camels, &c.). 8. That of elephants. 9. The order containing the dugong and manatee. 10. Whale sand porpoises. 11. The order of sloths, ant-eaters, and armadillos,— called the order of edentates. Such were all the orders any one recognised when the kangaroo was discovered by Cook and Banks. We must for our present purpose call special attention to Nos. 2, 4, 6, 7, and 11 of the above enumerated orders—namely, the orders of flesh-eating, insectivorous, gnawing and hoofed beasts, and the edentate order. Flesh-eating or carnivorous beasts have the teeth modified in a way of which those of the dog or cat may serve for types. Insectivorous beasts have the grinding teeth bristling with sharp points, well calculated to pierce the firm envelope of an insect's body. The rodents, or gnawers, have teeth like those of the squirrel or marmot—that is to say, very few in front, and these separated by an interspace from grinders placed more posteriorly which are adapted to crush vegetable food. Some squirrels are known as flying squirrels, and take long jumps with the aid of an extension of the skin of either flank. The hoofed beasts have long legs adapted for getting quickly over the ground with a reduced number of toes to each foot. Lastly, such creatures as the ant-eaters have no teeth at all, but have a long, worm-like tongue, and very powerful claws, while the sloths have feet especially modified for sure and slow progression in trees.

We may now turn to consider what the beasts were like which naturalists were astonished to find living in the fifth (Australian) "quarter of the world."

In the first place, there was the wombat, a burrowing creature with squat body and an extremely short tail, not unlike a marmot. The apparent affinity suggested by this external resemblance was confirmed when it was discovered that, like the marmot, and gnawing beasts generally, it had but a single pair of cutting teeth above and below at the front of the jaws, and that these were separated by a long interspace from the teeth well adapted for grinding vegetable substances. Here was a creature which might well be taken to be a true rodent.

It was originally described and figured in Col. Collins's account of the English colony of New South Wales in 1802. He tells us that Mr. Bass, when on an island in the straits named after him, "Bass's Straits," observed one of these animals walking with its usual shuffling gait. Having overtaken it, he placed his hands under its belly, and, suddenly lifting it, placed it on his arm with its back downward, as if it had been a child. "It made no noise," Col. Collins tells us, "nor any effort to escape us, not even a struggle." Its countenance was placid and undisturbed, and it seemed as contented as if it had been nursed by Mr. Bass from its infancy. He carried the beast upwards of a mile and often shifted him from arm to arm, sometimes laying him upon his shoulder, all of which he took in good part, until, being obliged to secure his legs while he went into a bush to get a specimen of a new wood, the creature's anger arose with the pinching of the twine, he whizzed with all his might, kicked and scratched most furiously, and snapped off a piece from the elbow of Mr. Bass's jacket with his grass-cutting teeth. Their friendship was here at an end, and

## THE OPOSSUM

the creature remained implacable all the way to the boat, ceasing to kick only when he was exhausted. In confinement wombats are mostly gentle. They will eat all kinds of vegetables, and are particularly fond of new hay.

FIG. 10.

THE VULPINE PHALANGER.

One observed by Sir Everard Home was attached to those who were kind to it, and would put up its fore-paws on their knees, and when taken up would sleep in the lap. It allowed children to pull and carry it about, and if it bit at all did not appear to do so from anger.

Allied to the marmot-like wombat were various tree-frequenting animals, termed phalangers, which feed on leaves, buds, and fruit, some of which naturally recalled to mind flying squirrels, as they were found to be aided in their long jumps by a similar extension of the skin of the flanks to that which exists in such squirrels. Here then we have another approximation on the part of these Australian beasts to the long familiar order of gnawers or rodents. We say "Australian," but although the first of these animals then discovered was found near Endeavour River, and named after Captain Cook, it was a closely allied form which Seba (as before mentioned) long before received from Amboyna, and regarded as the same animal as the American opossum.

The mistake was in those early days of zoölogical science by no means wonderful, for these phalangers presented some very remarkable resemblances to the opossum. Thus in both there was a pouch in the female, in both the tail was prehensile, and in both the hind foot was like a hand, with a well-developed opposable thumb. But if we look closely at the hind foot, we may detect yet another and yet more exceptional character. The two toes which come next after the thumb-like great toe are much smaller than the outer pair and more closely bound together by skin.

But another and very different set of beasts was also found in Australia. These soon gained the names of "native cats," "native devils," or "wolves," as the case might be. They are bloodthirsty flesh-eating animals, some of which have been compared to weasels and martens, and, indeed, such is the general resemblance of these creatures to members of the long-known order of ordinary carnivorous beasts, that there is little wonder that even Baron Cuvier placed them within it. Most of them

# THE OPOSSUM

agree with the wolf and fox in having no "great toe" at all, while the other four toes of the hind foot are of nearly equal size. The largest of the group only inhabits Van Diemen's Land, and is known as the "Tasmanian wolf." Some of its teeth are quite like those of the true wolf, to which it was naturally at first thought to be more or less closely allied. Thus, if the wombat and phalangers had some claim to be considered rodents, the creatures just noticed had at least as much claim to be included in the old order carnivora. Other beasts were

FIG. 11.

THE TASMANIAN WOLF.

found, however, which could advance similar claims to be associated with those before known insect-eating beasts, the insectivora, for, like the latter, their grinding teeth bristled with sharp points wherewith to pierce the hard skin in which most insects are encased. Some of these Australian insect-eaters are known as "phascogales," and are of the size of a rat, or yet smaller. Others are known as "bandicoots," and some of these exceed the hare in size. They are interesting, as we shall shortly see, because they have the hind limbs longer than the fore limb, and because the "great toe" is in them reduced to a small tubercle. In other respects also they present an exaggeration of characters before noticed as existing in the phalangers. Thus, their second and third toes are very minute and bound together by skin to the very claws, while the other two toes are exaggerated in size, especially the fourth toe.

We must now return to the consideration of those Australian mammals, the kangaroos—animals which most of our readers have probably seen in confinement, or else know by report to be most expert and prodigious leapers. Some of them are very large animals, as bulky as deer, and rapidity of locomotion is especially necessary for a large animal which inhabits a country subject to such severe and widely extended droughts as is Australia. The herbivorous hoofed beasts which were till recently so numerous in the plains of southern Africa—the antelopes—are also capable of very rapid locomotion. In the antelopes, however, as in all hoofed beasts, all the four limbs (front as well as hind) are exclusively used for locomotion. But in the kangaroos we have animals which require to use their front limbs for purposes of more or less delicate manipulation, with respect to the economy of the "pouch." Accordingly, for such creatures to be able to inhabit such a country, the hind limbs must by themselves answer the purpose of both the front and hind limbs of deer and antelopes. But the kangaroo's limbs are quite admirably suited to its needs. The front pair serve as prehensile manipulating organs, while the hind pair amply suffice to carry the animal over great distances and rapidly traverse wide, arid plains in pursuit of rare and distant water. This harmony between structure, habit, and climate was long ago pointed out by Sir Richard Owen.

## THE OPOSSUM

It might seem at first sight, then, that in the kangaroo we have a kind of creature allied to the hoofed beasts long before known, and only so far modified as to be in harmony with climatic needs. But the structure of its hind foot is alone sufficient to dispel such a notion. Each hind limb has, indeed (like that of an antelope or

FIG. 12.

THE KOALA.

deer), but two large and conspicuous toes. But these are of unequal size, and the inner one, which is much the larger, bears a very long and strong claw. On the inner side of this large toe is what at first sight appears to be a very minute one, furnished with two claws side by side. An examination of the bones of the foot, however, shows

D

that this apparently two-clawed toe, really consists of two very slender toes bound together in a common fold of skin. They are the toes that answer to the second and third toes of our own foot. Thus the kangaroo's hind foot, instead of being like that of the antelope, is a still further exaggeration of the foot of the bandicoot, just as that again is an exaggeration of the foot of the phalanger and the wombat. The really close relationship of these seemingly very different beasts is thus revealed.

A very distinct and very curious animal was also found in Australia, which passes the greater part of its life clinging to the branches of trees. It is, like the

FIG. 13.

THE ECHIDNA.

sloth of South America, slow in its movements, with a rounded head, long claws, a short body, and no tail. It is named the koala (Fig. 12), and eats the tender shoots of the blue gum-tree, feeding and sleeping at ease quite at the tops of the trees. It is very tenacious of life, and when even severely wounded will not quit its hold of the branch to which it may at the time be clinging. It is no wonder that this animal was often called by the colonists in Australia the native sloth.

Yet another strange animal was there discovered which

seemed to represent and belong to the before-named edentate order—that which contains the ant-eaters as well as the sloths. The animal we refer to is what we now call the echidna. It is a little larger than a hedgehog, and, like that animal, has its body protected by an investment of strong sharp-pointed spines. It has extremely powerful claws, and its long and slender jaws are entirely toothless, but contain a very long, extensile, and wormlike tongue. It is not surprising that it was called by Shaw at the end of the last century a spiny ant-eater.

"The French naturalists, MM. Quoy and Guimard, procured a specimen in Van Diemen's Land, which they kept alive for some time. They describe it as an apathetic and stupid animal, and state that for the first month after its capture it took no sustenance whatever, but at the end of that time it began to lap, and finally to eat some food prepared for it, consisting of a mixture of flour, water, and sugar. It avoided the light, remaining during the day partially rolled up with its head bent forward between its fore-legs. The rapidity with which it burrowed was astonishing. Being placed in a large cask full of earth, containing plants, it worked its way to the bottom in less than two minutes. Messrs. Bass and Flinders, who found one of these creatures, have related that their dogs could make no impression on it. It escaped from them by burrowing in the loose sand. It did not, however, do so head forward, but directly downward, thus presenting nothing but a prickly back to its adversaries as it descended. Another specimen was kept alive for some time by Lieut. Breton, who fed it on ants' eggs and milk. One kept in the Zoölogical Gardens in London was accustomed, when irritated, to roll itself up into a ball as a hedgehog does, the sharp points of its spikes then presenting themselves in all directions. When asleep it likewise rolled itself up. Of this specimen it was noted by Sir Richard Owen that its temperature was only 85° Fahrenheit, being nearly 10° lower than that of a rabbit. The resemblance of this animal to an ant-eater is increased by the fact that in order to keep its long tongue constantly supplied with a viscous substance (so that ants may adhere to it), it has enormous spittle glands (for its secretion), which extend from behind the eye to the fore part of the chest.

The females of all the animals from Australia and its vicinity which we have here noted are almost always provided, like the American opossum, with more or less of a pouch, and, whether they are so or not, they are all distinguished by the possession of two bones, called "marsupial bones," which extend forward in the flesh of the belly from the front margin of that bony girdle, the pelvis, to which the hind legs are articulated. We have said they exist in all. There is one exception: the Tasmanian wolf has this structure not in the condition of bone, but as two pieces of gristle, or cartilage.

Now, the possession of these marsupial bones or cartilages is found to go along with a variety of other characters which it would be out of place to enumerate here, but which serve to mark off the creatures possessing them in a very sharp and distinct manner from all other beasts. One very important character concerns the reproductive function and structures concerned therewith, and so all naturalists are now agreed that these Australian beasts, together with the opossum of America, form one great natural division which may be called "marsupial." All the orders of beasts known before, and above enumerated, form, on the other hand, a much larger and yet parallel group of animals which, from

## THE OPOSSUM 53

their mode of reproduction, are known as "non-marsupial," and "placental." But our short survey has shown us that the marsupial beasts are very diverse in structure. They are so much so that they contain groups which run parallel with various orders of placental beasts, as follows:

| PLACENTALS. | MARSUPIALS. |
|---|---|
| Rodents (marmot, &c.). | Wombat and its allies. |
| Carnivora (cats, weasels, wolf, &c.). | Native cats, weasels, and Tasmanian wolf. |
| Insect-eaters, mole, &c. | Phascogales and bandicoots. |
| Hoofed beasts (deer and antelopes, &c.). | Kangaroos. |
| Edentates (sloths and anteaters, &c.). | Native sloth and echidna. |

Now, according to the doctrine of evolution, all existing species are the descendants of common ancestors, from the structure of which they in various degrees diverge; and with regard to the origin of these two parallel series of marsupial and placental beasts, two hypotheses are open to us. One is that all beasts were at first of marsupial nature, and that the rodent, carnivorous, insectivorous, hoofed, and edentate placentals, are respectively the modified offspring and descendants of the rodent, carnivorous, insectivorous, kangaroo-like, and edentate marsupials. But on this hypothesis it is absolutely necessary that a number of very similar structures must be affirmed to have arisen independently. Such, in fact, must have been the case with regard to all those structural and functional characters by which the placental mammals agree to differ from all marsupials, since these characters must have similarly and separately arisen in each of these several groups, if we suppose the various groups of placentals to have severally descended independently from antecedent separate sets of marsupial forms.

The other hypothesis is that the whole of one group descended from one small section of the other, either all placentals from some one marsupial species, or all marsupials from one non-marsupial form. It is the latter hypothesis which is now in vogue, and the favourite opinion at present is that all marsupials descended from some insectivorous beast not very unlike a hedgehog, minus his spikes. But on this hypothesis, again, it is absolutely necessary that a number of very similar structures must be affirmed to have arisen independently. Such, in fact, must have been the case with all those structural and functional characters by which the various groups of marsupials, resemble the various parallel groups of placental beasts. Thus it is that the opossum and its allies exemplify "the independent origin of similar structures" more convincingly than almost any other order of beasts. They do so indeed in a way which makes denial simply impossible.

But among themselves alone, they force on our observation a subordinate instance of the same thing, and this is why we call attention to the various forms of structure presented by the marsupial hind-foot. We found its second and third toes becoming more and more bound together as we passed from the wombat, through the phalangers and bandicoots, to the kangaroos; while in the carnivorous forms, as also in the American opossum, these toes are as well developed and as independent as the others. We will call the former set, Group A., and the latter set, Group B. One of two alternatives, then, we must admit: either the forms contained in each group are specially connected by blood relationship and descent or they are not. If they are, then the various resemblances which may be detected between them, and which cannot be thought due to descent from a common ancestor, are

THE OPOSSUM 55

similar characters which must have arisen independently. If they are not so specially connected, then similar and extremely exceptional characters of the foot must have arisen independently.

FIG. 14.

THE CHÆROPUS.

But the structure of these curious feet teaches us yet another lesson. It cannot be supposed that the minute differences which exist between the second and third and fourth and fifth toes of the wombat are of vital importance to it, and yet they present us with what seems to be the initial, incipient condition of the character which becomes so very strongly marked in the kangaroo. It is more marked in the phalangers than in the wombat, but still we can hardly deem its condition such as to be of much, if any, importance to the life of those animals. But in the bandicoots it is so much marked that it may be of much use to them, and in the kangaroos it can hardly be otherwise. Nevertheless we cannot consider that the forms which possess this character in a slight degree are descendants of kangaroo-like creatures. Yet even if we did, we must, on the popular view of evolution, admit that the latter inherited it from antecedent unknown forms in which it was very slightly marked. It would seem, then, that here we have a character which has become gradually more and more developed, but at starting was of no appreciable use to the creatures possessing it. It would seem to have been developed for the service of other forms of life which were destined to come into existence at a later period.

But this character has been carried out to an even more exaggerated degree than in the kangaroo, and we will speak of the animal in which it is so exaggerated, and of one or two species besides, before we proceed to determine the precise position occupied by the American species, so that we may obtain a good answer to the question, "What is an opossum"?

The creature, the foot of which we have just referred to, is a very singular animal, which was discovered by Sir Thomas Mitchell on the banks of the Murray River. It was first described in 1838, and named *Chœropus* (Fig. 14). It is a slender-snouted, long-eared creature, less than twelve inches in length from the nose to the root of the tail, and with exceedingly slender and delicate legs and feet. Its

## THE OPOSSUM

food consists of insects and vegetable substances, and it forms a nest of leaves and grass. Its feet, however, constitute its main peculiarity. Every one knows that beasts differ somewhat as to the number of their toes—five to each foot being the maximum. In the horse, ass, and zebra alone are these reduced to a

Fig. 15.

THE MYRMECOBIUS.

single toe for each foot—those, namely, which answer to our middle finger and our own middle toe respectively. In sheep, oxen, deer, &c., each foot rests upon two toes only, but the chæropus walks upon six toes. Each of its fore-limbs is supported on two toes, while each of its hind-limbs, like each of the hind-limbs of the horse, rests upon one only. This single toe, however, is not the one the horse uses, but corresponds to our fourth toe, and to that which is the main support of the hind-limb of the kangaroo. Minute rudiments of the other toes exist beside it.

The next animal we wish to bring before our reader's attention is the *Myrmecobius*, an elegant, sharp-nosed, long-tailed, and bushy-tailed creature about the size of a squirrel, with the hinder part of the body ornamented with numerous transverse bands alternately light and dark (Fig. 15). It was first discovered by Lieut. Dale, in Western Australia, who found two specimens which had fled to hollow trees for refuge. The species was described and named by the late Mr. Waterhouse of the British Museum in 1836.

It runs with successive leaps, the tail being somewhat elevated, and every now and then will stop and raise its body, resting on its hind feet, thus altogether looking very like a squirrel. When caught it is harmless and tame, never attempting to bite, but uttering short, half-smothered grunts in its great alarm. The female appears to bring forth from five to nine young, using a hole in the ground or a hollow tree as a nest. She has no pouch, and the young are only protected by the very long and delicate hairs which clothe the region where the pouch is situated in other species. But the most remarkable character which the animal possesses consists in the great number of its grinding, or molar teeth, of which there are sixteen in the upper jaw and eighteen in the lower.

Now attractive as this little creature is, its interest for us consists in the fact of its being a "survival" of a very ancient state of things indeed. The opossum of America can lay claim to being of "old family," since it can prove its descent from the time when its relatives left their remains in the rocks beneath what is now Paris.

But that time was, after all, only in the tertiary period, and a "tertiary" family can have but a mushroom antiquity in the eyes of a creature which can establish its claim to have had ancestors in a flourishing condition during the secondary epoch. Yet this is just what the little myrmecobius can do. Its congeners even then lived in England, as is proved by the fact that their relics have been found in the Stonesfield oolitic rocks, the deposition of which is separated from that which gave rise to the Paris tertiary strata, by an abyss of past time which we cannot venture to express even in thousands of years. We have, then, in Australia what may be termed a surviving oolitic land, still showing us, in the present day, a living representative of forms which once indeed dwelt in the north, but have long since passed away from among us, leaving but rare and scattered relics "sealed within the iron hills."

When we pass from secondary and tertiary strata to deposits comparatively modern, we find that creatures closely allied to the kangaroo existed in Australia in times which must be called ancient historically, though very recent geologically. Just as in the recent deposits of South America we find the bones of large beasts, first cousins to the sloths and armadillos which exist there still, so in Australia there lived beasts having all the more essential structural characteristics of the kangaroo, yet of the bulk of the rhinoceros.

But, while we are speaking of fossils, we may mention an interesting circumstance which occurred with respect to the Paris predecessor of the American opossum found by Cuvier in the quarries of Montmartre. He first laid bare a lower jaw, and from a character it possessed—which is common to marsupial animals generally—he predicted that when the rest of the skeleton was uncovered, marsupial bones would be found present within it. Accordingly, in the presence of friends and admirers, he proceeded to remove the enveloping deposit with the greatest care, and so laid bare before the astonished eyes of his visitors, the very proof of the correctness of his prediction.

Quite recently there has been discovered in America a small mole-like beast which has been named *Notoryctes*. Some naturalists have considered it an Insectivore, but it is really a Marsupial, and a most interesting one, to which we shall again refer later on. Its marsupial bones are quite rudimentary.

FIG. 16.

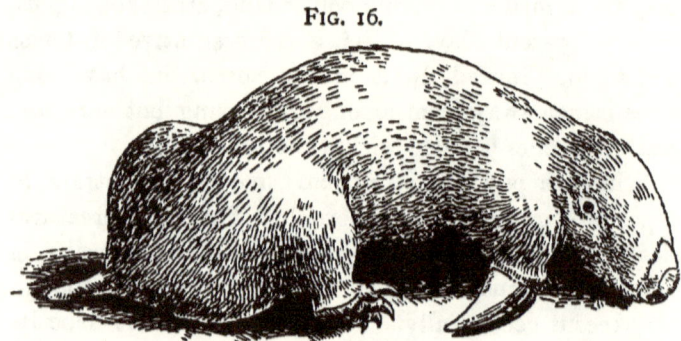

THE MARSUPIAL MOLE.

The last Australian animal we think it needful here to note is one which some of our readers may wonder we did not mention earlier. We refer to the *Ornithorhynchus* or duck-billed platypus, which gave rise, naturally enough, to so much astonishment when it was discovered at the close of the last century. Its bird-like mouth perhaps suggested the notion that it laid eggs, as a duck's bill is an awkward apparatus to suck milk with. It was found, however, that in the very young the bill is quite soft, and that it does at first feed on milk, like

# THE OPOSSUM

other beasts. Nevertheless, it is now certain that it truly lays eggs, and that the same is the case with the echidna.

FIG. 17.

THE ORNITHORHYNCHUS.

But these two animals differ from all other beasts in so many and such important points that they almost form a sort of zoölogical half-way house between birds and reptiles on the one side, and ordinary beasts upon the other. It is impossible here to describe these peculiarities; it must suffice to affirm their existence, and the reader who cares to pursue the subject further will find a description of them in every modern manual of comparative anatomy. These characters justify the separation of the platypus and echidna from all other beasts, and we must recognise that they form a sub-class by themselves, which, on account of the resemblance which in many respects it presents to birds, is called *Ornithodelphia*, and the platypus and echidna are termed "ornithodelphous mammals."

But the marsupials, apart from these two forms, are now also universally recognised as by themselves constituting another sub-class, which, on account of its uterine structure, is termed *Didelphia*, and all marsupials are, therefore, spoken of as "didelphous mammals."

All the rest of the class of beasts—*i.e.* of the class mammalia—constitute the third sub-class, which in the number of its species of course enormously exceeds that which contains the marsupials. This third sub-class, to which all those orders belong which were known before Australia was discovered, is distinguished as the sub-class *Monodelphia*, and all the creatures in it (the bat, the mole, the ape, the squirrel, the dog, the deer, the sloth, the ant-eater, the hedgehog, &c.), are known as "monodelphous mammals."

Our readers may now be able to appreciate how great was the hidden interest of that American beast known as the opossum. Little did those who first observed it suspect that it was an example of a group of animals so profoundly different from all mammals previously known. It was, in fact, impossible to appreciate its importance correctly till the beasts of Australia had been discovered and could be compared with other before known mammals and also with it. Moreover, though the American opossum is a marsupial, and possesses all the characters of the didelphous sub-class, it is none the less the representative of a very distinct family of that sub-class. It has been pointed out how distinct are the apes which inhabited the Old World from all those in America. It is just the same with the marsupials. There is no single species of marsupial found in Australia or anywhere else out of America which is in the present day also found in America. The American opossums are as much marsupials as are any marsupials. Nevertheless, they do not exhibit such great anomalies as do the kangaroos and bandicoots with respect to their feet, or the myrmecobius with respect to the number of its teeth.

But the geographical limits of the whole order, or sub-class, of marsupials are very interesting, for they are, at the same time, the limits of many other groups of animals and also of plants. We have not only an animal population (or fauna), but also a set of plants (or flora) which is characteristic of what is called the Australian region, that is extending not only over Australia and Tasmania, but more or less over New Guinea and the Moluccas, reaching as far north-west as the Island of Lombock, and even to Timor.

In India, the Malay Peninsula, and the great islands of the Indian Archipelago, we have another and a very different fauna and flora--that, namely, at the Indian region; and Indian forms of life extend downward south-east as far as the Island of Bali. Now, Bali is separated from Lombock by a strait of about fifteen miles broad. Yet that little channel is the boundary line between these two great regions—the Australian and the Indian. The Indian fauna advances to its western margin, while the Australian fauna stops short at its eastern margin. The zoölogical line of demarcation which thus passes through these straits is called "Wallace's line," because its discovery is due to the labours of that eminent naturalist and most persevering explorer. He showed that not only as regards beasts, but also as regards birds, these regions are thus sharply limited. Australia, he pointed out, has no woodpeckers and no pheasants, those widely spread Indian birds. Instead of these it has mound-making turkeys, honey-suckers, cockatoos, and brush-tongued lories, all of which are found nowhere else in the world. By becoming acquainted with all the various facts here detailed, it is possible to answer the question, "What is an opossum?"

We may say that an opossum is a form of marsupial life found only in America, and that it is a member of an order so peculiar as to constitute by itself a sub-class of the great class of beasts or mammalia. It is also a member of a sub-class intermediate between that to which the overwhelming majority of beasts belong, and the very restricted ornithodelphous sub-class which leads us down toward birds and reptiles. We may further say that the opossum is one important link in the evidence which, in some zoölogical respects, connects the South American continent with Australia, while at the same time it adds one more ground to the many which already exist for believing in the close zoölogical relationship between North America and the Europe of tertiary times. But the opossum has also another noteworthy distinction in that it exists in isolation in the midst of a vast continent which teams with non-marsupial forms of mammalian life. All the other marsupials live together in one mass, in all but complete isolation from non-marsupial beasts, the only exception being a few

# THE OPOSSUM

forms which are found in islands in the vicinity of Australia. There yet remains, with respect to the opossum, a most interesting question, as to which we are as yet quite unable to suggest any answer—the question namely, whence it came and from what line of ancestry. The answer to this question may not improbably be supplied by further discoveries among those fossil treasures which are rapidly making the North American continent the great centre of observation for all zoölogists interested in the past.

## III

## THE TURKEY

WHAT is a Turkey? This is a question on the consideration of which a little time may be spent, not unprofitably. There is no other bird which should indeed be so replete with interest for the English-speaking races of both sides of the Atlantic. Handsome in appearance, considerable in size, familiar as a *pièce de resistance* at family feasts of the highest political or deepest religious significance both in America and England, the turkey's scientific relations are also noteworthy, as we shall see presently.

But however interested Englishmen may be scientifically or gastronomically in the turkey, its claims on the appreciation and interests of Americans are, of course, greater. For almost every one now knows that the turkey is naturally an inhabitant of North America exclusively, and that no man has found one in a wild state anywhere else in the wide world. But there is more than this to be said as to its geographical exclusiveness. As we have seen, the Virginian opossum is an inhabitant of North America exclusively to day, though in earlier times it dwelt in Europe. But all the evidence there is goes to show that in Miocene times, as now, the turkey was an inhabitant of America only.

Not without reason did the great Franklin recommend the adoption of this peaceful, useful, ornamental and nutritious bird as a symbol of the great, peaceful, and prolific American republic, in preference to a creature so useless and destructive as an eagle—a kind of bird common to every quarter of the globe, and so hackneyed as a national symbol that nothing less than representing it with three heads would serve conspicuously to distinguish it from the single or double-headed eagles of European monarchies.

When North America was discovered, the turkey was distributed very widely east of the Rocky Mountains throughout what is now the United States, though as a wild bird its range in our days has become very much restricted. The Spaniards doubtless first brought it to Europe, and it probably owes its English name to the fact of its having first reached England in trading ships from the Levant. The first description we have of the turkey is that given by Oviedo in 1525, in his "History of the Indies." In 1566, however, twelve of these birds were presented to the French King Charles IX., and the first record of its appearance at a state banquet was at his wedding four years later. Soon after that it seems to have become common in England, and already to have found its place as a family dish at Christmas dinners.

We now propose to consider : (1) what are the turkey's nearest allies among birds ; (2) what relation the group to which it belongs bears to other groups of the class of birds ; (3) the relations which it bears as a bird to other animals.

The most anciently domesticated bird, and the one now most widely diffused, is, of course, the fowl so extensively bred by the ancient Egyptians and so brutally used in England down to the beginning of the last century. Besides cock-fighting, there were the much less known sports of "hen threshing" and "cock throwing."

For the former amusement a live hen was slung over a man's back who also carried horse bells, and was pursued about some court or enclosure by a number of blindfolded fellows, each of whom held a bough, with which they sought to kill the hen, amusing the bystanders meantime with the blows they bestowed on one another and on the bearer of the hen.

"Cock throwing" was a sport practised on Shrove Tuesday. The bird being tied by the leg, the thrower had to stand twenty-two yards off from it, and then try to knock down the bird by throwing a stick, after which he had to run up and catch it before it could recover its legs. In London, so late as 1680, money paid for this cruel sport was one of the sources set apart for the maintenance of the poor, and it was a recognised amusement of the lads of English public schools till about the year 1700.

The parent of the domestic fowl (*Gallus bankiva*)—the Bankiva fowl—is found wild over a very extensive range of country, usually from the Himalayas down to Timor and the Philippine Islands. It much resembles the game fowl, as also do several other species of the same great Indian region. To that region also belong the peacocks and various handsome pheasants. Peacocks are common enough in various parts of India, nor can those who have once enjoyed it easily forget the glorious sight of a number of these birds displaying their gorgeous plumage in the sun. But the Javan peacock, with its lovely neck of green and gold, is even more beautiful. We all talk of the "tail" of the peacock, and yet the feathers which form the part we thus name, are not really "tail feathers," but answer to the much smaller ones which cover the base of the ordinary tail feathers in most birds. They are what are technically called "tail coverts." The real tail feathers of a peacock are the

Fig. 18.

THE PEACOCK PHEASANT.

short and strong ones which stand up and support the magnificent plumes of the "peacock in his pride," as heralds term it.

Not less wonderful than the so-called tail of the peacock are the wings of the Argus pheasant. Though the body of the bird is not much larger than that of a fowl, the wings are each nearly three feet long, and so wide as to be of quite unwieldy size for flight, while the tail feathers are nearly a yard in extent. But it is the eye-like spots on its wings which are the most remarkable. A series of these, ornament each long wing feather, and when the large wings are expanded so that all their eyes are displayed (as they are in time of courtship), the effect is wonderful.

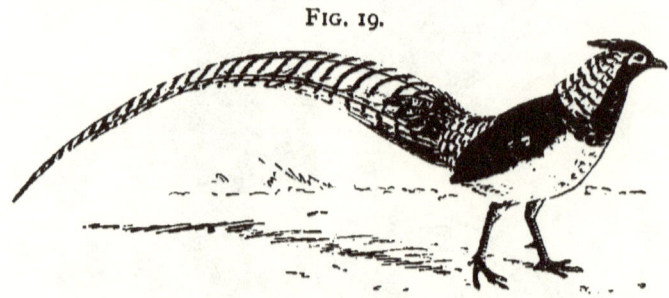

Fig. 19.

LADY AMHERST'S PHEASANT.

A smaller but still more beautiful bird is the peacock pheasant (Fig. 18), which has its tail as well as its wings covered with "eye spots."

The common pheasant never at any time was a really wild bird in the British islands, but was introduced into England at so early a date that it figured at feasts on the tables of our Saxon kings. Its true home seems to have been between the south of the Caspian and Black Seas. It is still wild in that vicinity, in the valleys of the Caucasus, the northern parts of Asia Minor, and in Corsica.

In Southern and Central Asia there are as many as thirty-five different kinds of pheasants, the most beautiful of which is Lady Amherst's pheasant, the plumes of the bird being like those of the gold pheasant in form, although its colours are far more delicate, harmonious, and refined. The longest tail of all is met with in Reeves's pheasant, the tail feathers of which bird may exceed seven feet.

Besides pheasants, certain curious birds called tragopans are found in India and Southern China. They are often spoken of as "horned," because of a soft piece of fleshy substance, shaped like a finger, which is attached to the side of the head on each side behind each eye. It is of different colours in different species of tragopans, and, though it ordinarily hangs down, it can be erected, when the bird really seems as if it had a pair of horns. A piece of distensible flesh also hangs down in front of the throat.

Now, all the birds (apart from the turkey) yet mentioned—namely, the fowls, peacocks, pheasants and tragopans—have long been recognised by naturalists as being birds near akin, and so they have been spoken of as *gallinaceous* birds, from the generic name *gallus*, long ago assigned to the most familiar kind, the fowl.

The birds just described are all Asiatic forms, but when we cross the Isthmus of Suez or the Red Sea into Africa, we bid adieu to every one of them; yet although we meet there with no peacock, fowl, or pheasant, we do meet with a small group of peculiar forms which are allied to them in nature, however different they may be in aspect. These African gallinaceous birds are the Guinea fowls, a variety of species of which range from northern to southern Africa and into the great Island of Madagascar. None of them approaches the pheasants in beauty of plumage or in grace of form, and every one knows the sobriety of tint of that Guinea fowl which has been introduced into our farm-yards. Yet some species, notably the vulturine Guinea fowl, are handsome and rather gaily coloured birds.

The vast region of Australia is entirely destitute of all pheasants, fowls, or peacocks, and also of Guinea fowls. Nevertheless we find there certain large birds which, though exceedingly exceptional as regards their habits, must be considered as allies of the above-mentioned species. These are the birds — called "mound-building" or "bush turkeys—which differ from all others as to the mode in which their eggs are hatched. Instead of sitting on them, they deposit them within large mounds of earth, which they heap up with their very powerful feet, wherein they also deposit more or less decaying organic substances. It is from the heat given forth by this decaying matter that the eggs of most of these birds are hatched, though some deposit their eggs in the sand of the seashore and there leave them to be hatched by the heat of the sun. They lay very large eggs, and the young within them become so matured and well-feathered before they are hatched that they are said to be able at once to fly away so soon as (after leaving the eggshell) they have found their way to the surface of the mound.

If we were to cross the Pacific from Australia, land on the coast of South America, and traverse the Andes to the forest regions of Brazil, we should there meet with yet other new and peculiar kinds of "gallinaceous birds," also of large size. There are the curassows, birds which may be seen in most well-organised zoölogical gardens, and are always to be found in those of London.

The curassows are sober-coloured birds, and (like so many Brazillian forms of life) more especially adapted for living in trees than are their allies of other regions.

# THE TURKEY

It is to this widespread group of gallinaceous birds (so differently represented in each great region of the earth's surface) that a new member was added when, North America being visited, the turkey was discovered. But the turkey of the United States does not stand alone; another species is found in Central America which is known as the ocellated, or Honduras turkey, and it is a most gorgeous bird. It is indeed one of the most gorgeous of all " gallinaceous birds," with tints of blue and

FIG. 20.

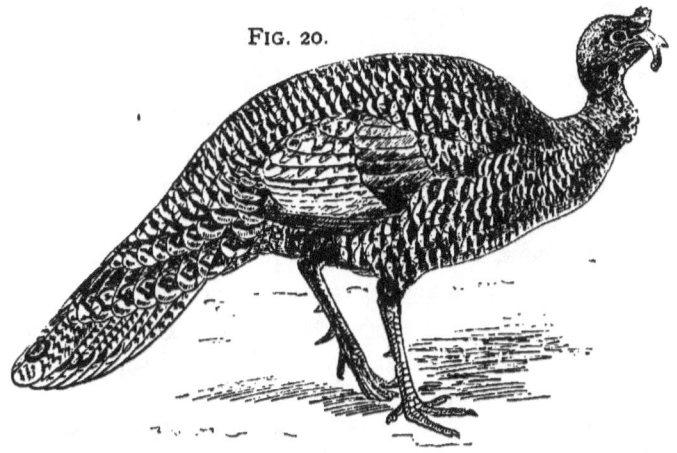

THE OCELLATED TURKEY.

green and red and gold, and beautiful and brilliant eye-like spots upon the tail feathers.

Such, then, is the turkey as regards its nearest ornithological allies. It, and the ocellated species, are the representatives in America north of the Isthmus of Panama, of the peacocks and pheasants of India, the Guinea fowls of Africa, the mound builders of Australia, and the curassows of South America.

These creatures, together with the partridges, the grouse and the quails, constitute the entire order of gallinaceous birds. We come now to our second question. What is the relation borne by their order (in which the turkey finds its place) to the other ordinal groups of the class of birds?

Now the class of birds is remarkable for the great number of species it contains, and the general uniformity of structure which they all possess. There are probably nearer twelve than ten thousand different species of birds, and in order that the human mind may be able to group in the imagination such a multitude of forms, an elaborate system of classification is evidently necessary. On the other hand, to arrange satisfactorily a multitude of forms, which are very much alike, is obviously a very difficult task. In the early days of ornithological science birds were roughly divided into birds of prey, perchers, scratchers, cooers, climbers, waders, runners, and swimmers. Of these eight primitive sets, it is the "scratchers" which answer to the gallinaceous group, and they have kept together; whereas the progress of knowledge has much divided and modified all the other sections of that primitive arrangement.

The subject of the classification of birds has greatly exercised naturalists of late years, but to deal adequately with that subject would require a long chapter. The characters made use of to distinguish the various ordinal groups also, are too technical and minute to be given here. We must limit ourselves to a list of types of such groups.

These will be: 1, An immense group of mostly small birds, from the crows and birds of paradise, to the humming birds; 2, the kingfishers and their alllies; 3, the woodpecker and its kin; 4, the cuckoos; 5, the doves; 6, the parrots; 7, the eagle and owls; 8, the pelicans; 9, the herons; 10, the bustards and rails; 11, the gallinaceous birds; 12, the snipes, &c.; 13, the gulls 14, the auks; 15, the ducks and geese; 16, the penguins; 17, the tinamous (birds of South America which approach the ostrich in the shape of the skull bones), and 18, the ostrich itself

FIG. 21.

THE RED BIRD OF PARADISE.

and its allies. The species here selected in each case are given merely as types; there are, of course, many other forms in each group which cannot be referred to separately.

The turkey, then, is a member of a small and peculiar

group of birds, different members of which have their home in different quarters of the globe, while the order contains the most ancient species ever domesticated by man.

But it is also one of a great number of species of the bird class, which are most peculiar in very different ways and yet all of them agree in being entirely confined to the American continent.

These birds are not—save the curassows—members of the turkey's order, and the only bond between it and them, is a geographical one—the fact that they are found nowhere in the world save in North or South America. It is the southern and central portions of the continent which contain the enormous majority of such forms, although most educated persons have heard of the mocking bird, the passenger pigeon, and the canvas-backed duck, as peculiar to North America.

To the south of that region we find the immense majority of species of those living gems, the humming birds, which it is the distinction of America to possess exclusively. The warmer parts of that continent also contain more species of parrots than are to be met with in any other quarter of the globe.

Even in the United States a species of parrot still lives in Florida, while eighty years ago it was abundant further north. Yet no members of its order exist in even the warmest corner of Europe. Not that a great warmth is an absolutely necessary condition, for a company of cockatoos long lived in a semi-wild state on a gentleman's property in Norfolk who had introduced them there, and had their food in winter carefully provided for them.

But the warmer parts of America are also the exclusive home of those singular and beautifully plumaged birds the toucans, which are so distinguished by the

## THE TURKEY

possession of beaks as remarkable for their extreme lightness as for their seemingly unwieldly size. There also alone are to be found tanagers, jacamars, motmots,* and bodies, besides some very singular and exceptional forms.

Fig. 22.

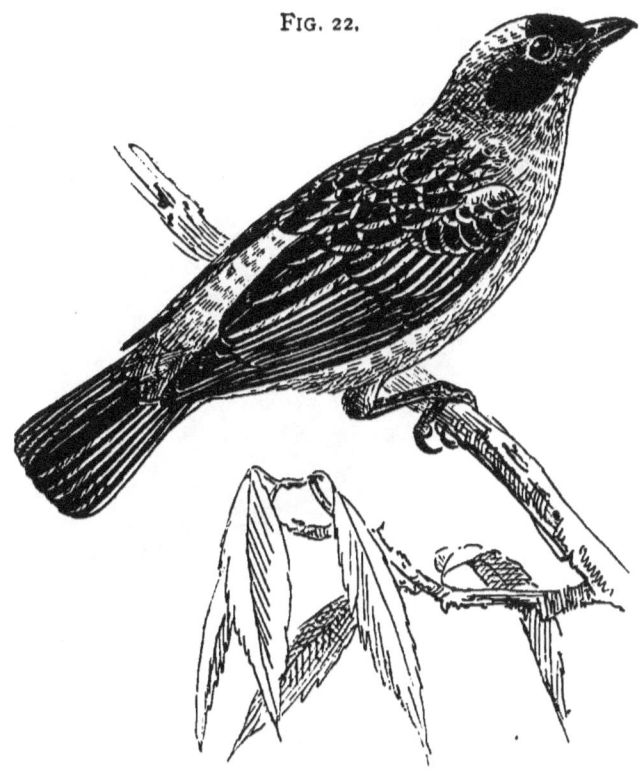

SCHRENCK'S TANAGER.

One of these latter is the horned screamer, which appears to be really a much-modified and arboreal goose, as does a still

* For notices and figures of these and many other peculiar birds, see the author's "Elements of Ornithology."—R. H. Porter, Princes Street, Cavendish Square.

more screaming, though hornless, form, called the chauna. Another is the cariama, a creature so puzzling that while some ornithologists regard it as a sort of hawk, others consider it a kind of bustard. There also are vultures such as

FIG. 23.

THE CHAUNA.

the huge condor of the Andes and the king vulture—both of which are so exceptional in structure that some of the most advanced of our naturalists altogether deny their claim to be considered vultures at all. South

# THE TURKEY 79

America has also its own ostrich-like bird, the rhea, which nevertheless presents us with a quite peculiar type of bird-life and has the bony girdle of its hip differently constructed from that of any other bird on the face of

FIG. 24.

THE CONDOR.

the earth. Another South American bird, one which lightly resembles a parrot in aspect, is known as the "Hoatzin," and is a very odd creature. Besides certain anatomical peculiarities it would be out of place here to notice, its juvenile condition is most exceptional. Then, part of its wing which answers to our hand is extraordinarily large for a bird, and is provided with two long claw-bearing fingers, by the help of which it is said to move about more like a quadruped than the bird which later on it proves itself unmistakably to be.

The turkey is thus one of a goodly company of peculiar American birds, but it remains for us to see what is implied in the fact of its being a bird at all. It is impossible to understand fully what a turkey is unless we know what are the characters it possesses in virtue of its avian nature, characters which it shares with other members of its class—the class of birds.

Now a bird is one of the most wonderfully organised of all animals, and almost the whole of its organisation is arranged for flight. The turkey is not particularly distinguished in this respect, nor are gallinaceous birds generally, and some of them, as the argus pheasant, seem to be quite exceptionally deficient in powerful and easy aërial locomotion. Nevertheless they all participate in those essential peculiarities which facilitate the rapid gyrations and the wide wanderings of their more capable cousins. The flights some birds will take are indeed amazing. A falcon which belonged to Henry IV. of France, is known to have flown from Fontainebleau to Malta, a distance of 1350 miles, in one day. The celebrated race horse named Eclipse would run at the rate of a mile a minute, yet there are hawks which fly at a pace of 150 miles an hour. But not only the rapidity, but also the endurance which birds possess is wonderful. Swallows will migrate from England to the south of Africa.

Oceanic birds are, as might be expected, exceptional even among their class for long-continued flight. Many of them rarely come to land except to breed. The well-known stormy petrels, called by English sailors "Mother Carey's chickens," are not much larger than swallows, yet they will accompany a ship on its course for many days. Every one has heard of the albatross, and the frigate bird is also widely known. Both of these birds possess wings of enormous length, and the former is celebrated for its power of sailing in the breeze without once flapping its wings, till the observer is tired of watching it.

But the mode of life of a great number of birds forces them to execute the most extensive journeys. They have a persistent habit of departing toward the approach of winter from their colder quarters to seek warmth and more abundant food in other climes. This is "migration," and migrating birds always breed in the coldest parts they visit, whether in the northern or in the southern hemisphere. Little, however, is known of migration in the more southern regions of the globe, and the Antarctic lands are so extremely cold that they are not visited as are the lands toward the extreme north.

When the time of the autumn migration arrives, it is the young birds which, in spite of their inexperience of the route, set out first, save that an old bird will sometimes take it into its head to start before the regular time comes, but on the return journey the youngsters generally lag behind.

Thus locomotion of an extraordinary kind is habitual with birds—locomotion both rapid and prolonged, and effected by persistent reiterated strokes of the wings against the resistance which the atmosphere opposes to their efforts. These efforts are also made by the help of vigorous bones and muscles, and a flying bird falls immediately when any injury has paralysed its muscular power, as every sportsman knows.

That such feats of flight should be performed by a creature so solidly built needs a very careful and a very peculiar arrangement of the parts and organs which compose its body. The various organs are indeed so packed and arranged as to make the centre of the body's gravity fall just where it can be best sustained, and they are so constructed as to produce, in combination, the greatest strength and warmth with the least possible weight. But no dissection is needed in order that we may see how admirable a bird's organisation is. Its very external clothing shows us this, for it is made up of feathers, and a feather is a very marvel of skilful construction. It is a structure at once so light and so strong, so admirably adapted to retain the heat of the body that it is hardly possible to imagine anything of the kind more perfect.

A bird's bones must be strong for the work they have to perform, and to be strong they must possess a certain solidity, and therefore weight, but this weight is commonly reduced to the minimum necessary for the needful strength. This reduction is effected partly by a skilful arrangement of the solid parts themselves, but it is further effected by the bones containing not a mass of marrow, but warm (and therefore light) air. In some birds even each bone of the toes is thus furnished, while air passes into the bones of the skeleton freely by means of passages, which extend to them from the lungs. But in order that a bird may be able to fly, great power is no less indispensable than is lightness of structure, and to move the wings as they need to be moved for flight, very large muscles are necessary. Large muscles are necessary to raise as well as to depress the wings

## THE TURKEY 83

rapidly. Such opposite actions are produced in other animals by muscles placed on opposite sides of the body, but it would never do for birds to have a heavy mass of flesh on their backs. Accordingly, by a special and most simple contrivance, both movements are brought about by muscles conveniently placed in layers on the under surface of the body. The muscles which pull down the wing act directly on the wing bones, but the muscles which pull them up, act indirectly: the sort of cord in which they end being bent round a bony pulley so as completely to reverse the action they would otherwise produce. This is the reason why there is so much meat on the breast of a bird—such as a partridge or a woodcock. It consists of the muscles which both pull down the wings and raise them. The relation which exists between the volume of these muscles and the power of flight is well instanced by the difference in the quantity of meat we find on the breast of a wild duck and a tame duck respectively.

A bird might, however, have the most voluminous of muscles, but they would be of small use to it, were they not stimulated by a copious supply of vivifying blood, kept pure by a most efficient process of respiration. And these aids are supplied with very exceptional completeness.

It is plain that in such a creature as a bird, the further any organ may be removed from the centre of gravity, the more necessary it becomes that it should be neither heavy nor bulky. Thus a bird's arm and hand are reduced to what is just necessary to sustain and wield the large feathers which form the wing. The hand is especially diminished, and its few rudimentary fingers are closely bound together in a fold of skin. And yet birds have a great many "handy" actions to perform. What dexterous " manipulation " is needed to construct the wonderfully delicate, yet strongly woven nests so many of them form! But as they have no hand to use, and cannot (as monkeys can) use their feet as hands, they are forced to manage another way. And they do manage admirably with the help of their bill, which is indeed a " handy " organ and serves as a most skilful and delicate instrument for prehension. But birds could not thus use it if their neck was as little mobile as is our own. Their neck, therefore, is very mobile—so mobile that, even when it is short, the head can be turned round so as to look directly over the back. The exclusive devotion of the arms and hands to flight, makes it necessary for a bird to devote its other limbs entirely, or almost entirely, to locomotion. For this purpose the legs need to be supplied with vigorous muscles, and yet every one must have remarked what thin and delicate legs birds have. The fact is, they have very strong and voluminous leg muscles, but then they are gathered together in the thighs and upper parts of the legs—near the centre of gravity. But though thus distant from the toes, they are enabled to act upon them by the aid of long and delicate cords, or tendons, which run down side by side through the thin parts of the legs to end in the toes, on which (by the insertion of their tendons) the strong leg muscles act. So complete, indeed, is the process of packing everything as centrally as possible carried out, that birds even carry their teeth in their stomachs instead of their mouths, although Prof. Marsh has shown us that in the "good old times" there were birds who held to the ways of their probable forefathers and kept their teeth in their jaws. But they are not real teeth which our modern birds bear in their breasts, though they answer the purposes of true teeth  They are small stones, which very many birds swallow. This is especially the habit of birds which feed on grain and other similar substances, which are ground and comminuted by the gizzard with the help of the stones it is made thus to contain. Again, we all know that when we sing we make use of our larynx, an organ situated in quite the upper part of our throat, close beneath the root of the tongue. We all also know how beautifully many birds sing, and we are sure, without being told, that they must have their organ of voice also. Such an organ, indeed, they have, called the "syrinx." But this organ of voice is placed at the bottom of their throat instead of at the top; it is situated, again, as nearly as it conveniently can be to the centre of gravity.

A very wonderful organ is the eye of a bird. To say that a man has "the eye of a hawk," is highly to compliment his power of vision. But few persons who use that expression realise how great a eulogy it expresses. A rapacious bird, such as a hawk, has to keep a close eye upon prey which may be running about on the ground while it is watched by the hawk from a great altitude. When the hawk "stoops," that is, pounces on its quarry, it is necessary for it to keep its victim well in sight during the whole of the hawk's rapid descent from so great a height. How delicate and extensive must be the power of adjustment which a hawk's eye possesses in order to enable it to effect this!

The eye of a bird needs to be kept very clear and bright, and birds possess a special mechanism for sweeping the eye rapidly and often. If we watch a hawk we may observe that its eye frequently becomes shrouded for a moment by some delicate film passing over it, which is, in fact, a third eyelid. We seem to have nothing like it, but we really have the rudiment of such a structure ourselves. At the inner angle of every human eye there is a minute fold of skin which serves no known purpose whatever, but is the representative of the bird's third eyelid or "nictitating" (that is winking) membrane. A truly wonderful mechanism also exists in connection with this membrane. It is drawn over the eye by the aid of a muscle with a delicate tendon, which sweeps round the nerve of sight (the optic nerve), and would injuriously compress it were it not that, on its way, it passes through a loop-like tendon belonging to a distinct muscle, which, acting at the same time, pulls it sufficiently away from the nerve of sight to avoid all ill effects.

The turkey shares in all the above-noted peculiarities of bird structure and bird habits, nest making, and careful care of the young among the latter. Its nest, however, is a very poor affair compared with that of the majority of land birds, and consists but of a few dried leaves or twigs on the ground, perhaps under the shelter of some bushes or of a fallen tree. It is the weaver birds—birds of the Old World—which construct the most elaborate nests known. They construct immense ones, or rather a huge cluster of nests placed sociably side by side, under one cover, each nest having its own separate entrance on the under side of the whole structure, and not communicating with the nest next to it. The whole mass may be ten feet in diameter, and ultimately break down from its own weight.

For details as to the habits of the turkey, readers may be referred to American ornithologists, one of the most distinguished of whom is Dr. Elliott Coues. But Audubon long ago gave a graphic picture of the parental care of the female turkey. It is the Old World, however, which affords us the most striking examples both of parental and conjugal virtue and defect. There are to be found the most immoral cuckoos as well as those most virtuous of birds, the hornbills. The hornbills inhabit both Africa and India, and have beaks almost as large, relatively, as those of the toucans, and much more hard and dense in structure. The hornbill is a large bird, and makes its nest in a hollow tree, and when the female has taken up her station within it, her thoughtful mate forthwith proceeds to imprison her, closing up the mouth of her retreat by means of a partition of mud. The kindness of this action may at first seem questionable, but it is not really so. In the first place the husband is careful to leave a small aperture in the partition, of which aperture he afterward makes a most exemplary use. Leaving his wife to pursue uninterruptedly her maternal duties he forthwith devotes himself most zealously to her support, wandering almost incessantly about in search of food, and returning to her again and again to minister to her needs, and to feed her through the small opening left for that purpose. So great is his devotion to this conjugal duty that by the time his progeny come forth from their enclosure, their sire may have reduced himself to the most sorry plight, sometimes even falling a victim to the exhaustion brought on by his devotion to the needs of his nesting spouse.

But to return once more to the turkey, it may be said, with the other members of the order of gallinaceous birds, to occupy a medium position in the whole class. Neither in structure nor in habit does its order show extreme peculiarities, save and except the tail of the peacock, the wings of the argus pheasant, and the exceptional egg-laying habits of the mound builders. In its own order, however, the turkey may claim a distinguished place from its utility, its size, and the gorgeous beauty of the ocellated or Honduras species.

Such, then, is the turkey in itself, a member of the gallinaceous order, and such are the structural characters it shares with other birds. It now only remains to point out the leading characters whereby it, with the other members of its class, differs from creatures which are not birds. The class of birds, as one of the classes of back-boned animals, stands between the class of beasts on the one hand and the class of reptiles on the other.

We saw, when considering the opossum, that the class of beasts is divided into a number of very different orders. How greatly these differ from one another will be apparent if we recall to mind the obvious dissimilarity which exists between an ape and an ox, a bat and a lion, a mole and a squirrel, a seal and a mouse, an elephant and an armadillo, or an antelope and a whale; and yet these are all of them beasts. We also saw how great a diversity may exist in a single order of beasts—such as that, *e.g.*, to which the opossum belongs.

Not less striking are the contrasts and divergences which exist among the various kinds of reptiles, such as between serpents and alligators or between terrapins and lizards.

If, however, we take into account the forms of reptilian life which have passed away since the deposition of the chalk cliffs of the English coast—since, that is, the end of the secondary period—the contrasts and divergences become yet more striking.

In those early times, instead of porpoises and whales, the sea swarmed with reptilian predecessors of those short-necked beasts. It swarmed with ichthyosauri, often of great bulk, and also abounded with those large aquatic creatures with more than swan-like necks, the plesiosauri. Huge reptiles also grazed in the then existing fields or fed on the leaves, fruit, and twigs of forests.

These vegetable feeders, again, were preyed on by fell monsters compared with which our lions and tigers seem insignificant. The air also was agitated ceaselessly by the wings of flying reptiles (pterodactyles) of all sizes, flitting like bats, but sometimes with the proportions of the albatross.

This hasty glance at the two classes of beasts and reptiles will enable us to appreciate the distinctness of the class of birds. They are the most easily defined of any class, since the single epithet "feathered" suffices to characterise them. It does so because every bird has feathers, while no such thing is possessed by any creature which is not a bird.

Birds, as we have already said, stand between beasts and reptiles, but are widely distinct from them both. All beasts possess, as we possess, warm blood, but the blood of the bird is warmer still, and thus birds differ greatly from reptiles, in spite of their possessing certain structural characters in common with them. For the reptile the blood is hardly ever warmer than the medium which may surround it, the only exception known to us being that which occurs in a boa-constrictor or python, when hatching its eggs, around which it twists itself in a conical coil, surmounted by the ever-watchful head.

But birds differ from both beasts and reptiles, by that singular uniformity of their structure before adverted to. Some beasts and reptiles have but a single pair of limbs. Thus the whales and porpoises have but a single pair each, and some reptiles—serpents and some lizards—have none, but every bird has two pairs. The limbs of beasts and reptiles may be constructed variously. How different are the wings of the bat from the scoop of the mole, the paddle of the whale, and the foot of the horse. But in birds, the hind limbs are always " walking " legs, and fore limbs " wings," although, as in the ostrich and the apteryx, they may be incapable of flight. They are always, that is to say, formed on the type of the flying wing, even when they cannot sustain or propel the body in the air. We see this emphatically in the penguin, a bird incapable of so rising, but a most wonderful swimmer beneath the surface of the sea. There it propels itself by the aid of its fore limbs, which are worked powerfully by enormous muscles, so that the bird may be said to fly under water. Very many beasts and reptiles have long tails, and some have none, but every existing bird has its tail feathers supported by a short structure of flesh and bone.

Most beasts have a hairy coat, and a few are naked, while some, as the pangolin and armadillo, are clothed with scales or bones, and much diversity exists in the covering of reptiles. But all birds are, as before said, clad with feathers as to their body and with scales as to their feet. Almost all beasts and reptiles have teeth in their mouths, though a few, like the ant-eaters, turtles, and terrapins, have none. But no existing bird has teeth, while all have their jaws clothed with that horny investment we call the bill.

We have, however, compared the class of birds with extinct as well as with living reptiles, and, therefore, we cannot fully comprehend what the class of birds is, nor what the turkey's profoundest relations to other creatures are, if we take no account whatever of birds which have passed away. And very many a bird has here and there passed away which was known to a few preceding generations of mankind, if not to our own fathers.

Thus, that handsome bird, the great bustard, was not so long ago abundant in the more open parts of England

-such as Salisbury Plain and the Sussex Downs, and it lingered on in Yorkshire down to 1830. The spoonbill disappeared earlier, having been exterminated toward the end of the seventeenth century.

A very interesting bird, once an inhabitant of Britain, which has now become utterly extinct everywhere, is the great auk. It was of about the size of a goose, but, having very small wings, was quite unable to fly. Being a powerful bird, with a strong bill, and a most accomplished diver, it would have continued to live on in full security but for man's reckless destruction of it. Unable to rise in the air and deposit its eggs in security on high ledges of rock, it could only shuffle along some gentle slope to lay its eggs at a safe distance above high water mark. It was formerly very abundant, and hundreds of auks at a time were taken off the coast of Newfoundland in the first quarter of the seventeenth century.

In 1813 the auk was abundant on the rocky islands off the coast of Iceland, and in 1844 two auks were caught on Eldey Island. Since then the auk has disappeared altogether from Europe and now has vanished from the whole world. About seventy-six skins and nine skeletons, with sixty-eight eggs and a few bones, preserved in collections, are all the relics we have of this now extinct and most interesting species.

The Labrador duck is another bird which has disappeared yet more recently, as it lived on till 1852. A curious and handsome starling (*Fregilupus varius*) has also disappeared from Mauritius, a most precious skin of which has been recently acquired by the British Museum (Fig. 25). The name of this island will remind many of our readers of its celebrated former inhabitant, the dodo, which became extinct by the end of the seventeenth century. An extinct bird of Madagascar has been named *Æpiornis*, and its egg is the largest one known. If, as is possible, its eggs were objects of commerce in ancient times, it

FIG. 25.

THE EXTINCT STARLING.

may be that we owe to it the fable of the "roc" (with which every reader of the Arabian Nights must be familiar), since it is a very natural error to suppose that the size of a bird must correspond proportionally with that of its egg.

But it is New Zealand which is justly the most celebrated country for its extinct birds. Till man visited it no beasts save bats dwelt there, and those gigantic birds, the moas (*Dinornis*), lorded it over all other living creatures, and stalked about in absolute security, without the power or need of flight, till man came and exterminated them. There also was once to be found a large bird of prey, known as *Harpagornis*, which had more powerful claws than any existing eagle possesses.

But all the extinct species yet mentioned must be reckoned as relatively modern forms of life. When, however, we descend the stream of time and explore the rocks deposited in past geological ages, we find unmistakable evidence that birds once existed which were very different indeed from any of those which people the surface of our planet in the present day. Nevertheless, as we recede we find the change which has taken place to have been a gradual process of change. In the Pliocene rocks we meet only with genera which now exist, nor are we struck with any marked geographical changes. But in Miocene times, trogons and parrots dwelt in Europe, but not the turkey, which even then seems (as before said) to have been an exclusive inhabitant of America.

When we penetrate into the Eocene rocks, however, we find genera altogether new, although allied to such birds as larks, kingfishers, vultures, woodpeckers, &c. The exploration of the secondary rocks, has, as might have been expected, brought much stranger forms to light.

The remains of a bird named *Hesperornis* have there been found which had true teeth in its jaws, but was yet more remarkable for what it had not. It had only a feeble upper arm bone, to the end of this was attached but a rudiment of the lower arm. There also has been found another bird with teeth, named *Ichthyornis*, because the form of the segments of its backbone recalled to mind the form of the same parts in a fish.

But the most ancient bird yet known was found in 1861 in oölitic strata in Bavaria. This was the renowned *Archeopteryx*, which differed remarkably from every other bird we know. We have already said that every existing bird, whether its tail feathers are long or short, has them supported on a fleshy pad, which contains the bones of a very short tail. This "pad" is what is known in the fowl as the "parson's nose." But the archeopteryx had no such short tail, but instead, a very long one, composed of no less than twenty bones, to each of which two long feathers, one on either side, were attached. Its hand also was very exceptional.

The changes which we thus see to have taken place in the course of ages, lend additional interest to the question—Whence did birds arise? It has been suggested that they were derived from certain long extinct reptiles. These reptiles have been named *Dinosauria*, and are represented by the large *Iguanodon* (discovered many years ago by Dr. Mantel), which once wandered and grazed in the Weald of Kent and Sussex and in what is now the Isle of Wight. Travellers to Europe may see a magnificent example of this creature, in the form of its skeleton, admirably set up, in the Museum of Natural History at Brussels. It has also been suggested that birds are allied, by descent, to those flying reptiles, the pterodactyles, and an in-

## THE TURKEY

genious cast of the inner side of the skull of one of these has shown that their brain was exceedingly bird-like. That the iguanodon-like reptiles were in some respects like the ostrich and its congeners is not to be denied; but then the ostrich and its allies are not creatures on the road to become flying birds, but seem rather to be degraded descendants of birds which once flew. Moreover, the oldest known bird, the archeopteryx, is not at all ostrich-like, but has much more affinity with ordinary birds, save as regards its hand and tail. Thus the origin of birds is a question still open to dispute, and while welcoming gladly light from any side upon the problem, we would carefully eschew a hasty dogmatism on that, as on every other subject.

# IV

## THE BULLFROG

WE have selected the bullfrog as one type of animal life in order to introduce to our readers' notice a group of animals about which we have been hitherto silent. By the aid of the ape and the opossum we have taken a preliminary survey of the two great groups (placentals and non-placentals) which make up the class of beasts. The turkey has aided us to portray the general characters of the class of birds, with a side glance at that of reptiles; a group which will be here glanced at once more and then reserved for fuller treatment hereafter.

But there is another group of animals, allied to fishes, about which we have been silent, and it is this one to which the bullfrog belongs. It is a group of animals which, we think, must be held to constitute a class by itself—a class containing creatures which seem very different externally, but are none the less fundamentally alike. This class, the frog's class, is sometimes called the class *Batrachia* (from the Greek word for a frog), and sometimes *Amphibia*, from the life experiences (as to breathing) which most of its members go through. We shall elect the former term and speak of the members of the frog class as "batrachians."

Just as we have seen the classes of beasts and birds to be each made up of orders, so also is the batrachian class

# THE BULLFROG

made up of orders, and these are four in number. Three of them are represented by living species, and the fourth by others which have been extinct for an unimaginable abyss of time, namely, since the deposition of the coal measures.

Our present endeavour will be, first, to mention certain facts about frogs and the whole of that batrachian

Fig. 26.

THE BULLFROG.

order of which they form part; secondly, briefly to describe the other three orders; and, thirdly and finally, to consider the relations which exist between the frog class and the other classes of backboned animals. For, just as we found birds to stand between beasts and reptiles, so we shall find that batrachians stand between reptiles and fishes, and also that as they advance in life they become less like the latter and grow more like the former.

But the frog has special claims on our gratitude and

commiseration on account of all it has done and suffered to increase our knowledge. In every physiological laboratory frogs are such ceaseless subjects of experiment that the animal may well be called the "martyr of science." What their legs can do without their bodies, what their bodies can do without their heads, what their arms can do without either head or trunk, what is the effect of the removal of their brains, how they can manage without their eyes, what effects result from all kinds of local irritations, from chokings, from poisonings, from mutilations the most varied? These are questions again and again answered practically for the instruction of youth, while the most delicate and complex researches are carried on through their aid by the very first physiologists of Europe. We know by the unhappy instances of men who have had their backs broken, that unmistakable results may be produced through irritations which are entirely unfelt. A patient may be induced to withdraw his foot when its sole is tickled with a feather, though he be utterly unconscious of both the tickling and the motion of his foot which it induces. Following up the indication thus given, it has been found that a frog, the head of which has been cut off, will raise one of its feet to rub a spot purposely irritated by some corrosive fluid, and that when the foot so raised is held or cut off, that then the other foot will be applied, instead, to the irritated surface. It is thence concluded, as a matter of course, that, the head being removed, the frog can no longer feel. We do not for a moment believe that it can feel, but we are bound to affirm that it is not evident to us that it cannot do so. We know nothing about even our own sensations except through the conscious intellect which accompanies our experience of them; what our mere "feeling" apart from that accompaniment may *in itself* be, we can only conjecture. In a creature with so small a brain as that of a frog, we cannot dogmatically affirm that the spinal marrow may not be an organ of feeling, although there is nothing to show us that it is.

Every one knows the soft, smooth, moist skin of this animal. Its skin is one of its most important organs. Indeed, our own skin is by no means popularly credited with the great importance really due to it. " Only the skin!" is an exclamation not unfrequently heard, and wonder is felt very often when death supervenes after a burn which has injured but a comparatively small surface of the body. Our skin is indeed a most important structure, and able, in a very slight degree, to supplement the action of the lungs as well as of the kidneys. In the frog it is really an organ of breathing, almost, if not quite, as indispensable as the lungs. Neither will suffice without the other. A frog may be strangely choked in two ways. To distend its lungs it is compelled to swallow air after closing its lips upon a mouthful of it. Thus a frog may be choked by keeping its mouth open. Again, no breathing (that is, no exchange of certain gases) can take place except on a surface which is moist; therefore, that a frog may breathe with its skin, that skin must be moist, and it is kept so by the exceptional ease with which water exudes forth from the body upon it. In fact Count Smalltalk only made Mrs. Leo Hunter speak accurately when he misrepresents her ode as being addressed to the "perspiring frog"—for the frog is one of the most perspiring of all animals. It is so to such a degree that one tied where it cannot escape the scorching rays of a summer's sun, will not only die, but soon become perfectly dried up—as we recollect discovering when a child, to our great sorrow and disappointment.

There are a very large number of species of frogs and toads. At least fourteen kinds inhabit Europe and North Asia and Africa north of the Sahara; above ninety are found in Africa south of the Sahara, some hundred and sixty in the Indian region, more than seventy-five in Australia, no less than three hundred and five in tropical America, and fifty-three in the North American region.

The phenomena of the life-history of some species of frogs and toads are very curious. The ordinary course of a frog's development takes place thus: The approach of spring calls them forth from their winter retreat, which is generally in mud under water. Great numbers of them are often dug up in the winter time, all clustered together, in the mud at the bottom of a pond. In the month of March their well-known croaking begins to be heard in England, and. though itself unmelodious, it possesses a certain charm through its connection with the vernal outburst of Nature. It is then that they congregate for egg-laying. Their eggs are little dark, round bodies, enclosed in no solid shell, but only in a thin glutinous envelope. The latter quickly swells in the water, so much so that the "spawn" in the case of the common frog soon comes to have the appearance of a great mass of jelly, through which dark specks (the yolks of the eggs) are scattered. By degrees each little dark mass assumes the form of a young tadpole, which emerges from the egg toward the end of April. At first it has long filamentary processes of skin projecting from either side of the neck, and these are the primitive gills or aquatic breathing organs. They soon become absorbed and are replaced by other shorter gills, which do not project visibly from the neck. Little by little the limbs bud forth and grow, and at the same time the tail is absorbed,* while apertures on either side of the neck close up, which were the external openings of the chamber in which the secondary gills lie, and the young frog then breathes by means of its lungs in the ordinary way. The tadpole is extremely unlike the frog it is to grow into. Not only does it breathe by gills in water, instead of by lungs in air, but at first it has a very long tail, with which it swims, and no limbs; while when a frog, it has no tail but long limbs, which are its only locomotive organs. The tadpole has a very small mouth and very long intestine, and feeds on vegetable substances. The frog has a very large mouth and very short intestine, and feeds only on animal matter.

The common frog is distributed widely over the Old World, though unknown to America, which, however, possesses another species very like it. Similarly, the bullfrog is unknown in the Old World, save in zoölogical gardens, where it is always welcomed as a curiosity.

The genus (*Rana*) to which these and other true frogs belong has its headquarters in the East Indies and in Africa, but extends over all the great regions of the world except Australia and New Zealand. In South America, however, there are but five species, while there are no less than fifteen kinds in North America.

The common toad (*Bufo vulgaris*) is as widely distributed as is the common frog. Three species of the genus *Bufo* are found in the northern parts of the Old World, seven in North America, and seven in Africa; twenty-two in the Indian region, and thirty-six in tropical America, but none in Australia. They all differ from frogs in being toothless, while the frog

* Thus, when Dickens makes one of his characters exclaim: "What next? as the tadpole said when his tail dropped off," he was more amusing than accurate.

has a row of teeth along the margin of the upper jaw. The toad has an oblong prominence behind each eye, from which a milky fluid exudes, which is very unpleasant to dogs, as they show by a copious flow of saliva and many headshakings if they happen to have seized a toad and have so come to taste this secretion. It exercises a very decided effect upon certain animals, since the tadpoles of frogs are very powerfully affected by being kept in the same water as a toad, if the latter be irritated and so made to discharge this fluid.

A frog (*Pelobates fuscus*) which is common in France and quite harmless, has, nevertheless, a singular provision for self-defence. If it be seized or its leg pinched, two effects follow: It utters a sound like the mewing of a cat, and emits a vapour which smells of garlic strongly enough to make the eyes water, as mustard or horse-radish will do.

This vapour and the toad's secretion are the nearest approach to venomous products which any members of the frog's order possess.

A small European frog, named *Alytes obstetricans*, has a very curious habit in connection with the regular life-history of the species. The female lays her eggs so that they adhere together in the form of a long chain. The male then twines this chaplet of his wife's eggs round and round his thighs till he acquires the aspect of a gentleman of the court of the time of James I. arrayed in puffed breeches. After having thus encumbered himself, he retires, at least during the day, in some burrow, till the period arrives when the young are ripe for quitting the egg. Then he seeks the water, into which he has not long plunged when the young burst forth and swim away, after which he makes himself tidy (frees himself from the remains of the eggs) and resumes his normal appearance.

Certain frogs are termed tree-frogs, and of their typical genus (*Hyla*) there are thirteen species in North America, and eighty-seven in tropical America, while only one has a home in Europe. These tree-frogs are remarkable for their adaptation to arboreal life, the ends of their fingers being spread out so as to form suckers, by which they can easily adhere to the leaves of trees. The European species, the green frog, has a wide range, though it does not extend to the British Isles. It is a very elegant, attractive little animal, but visitors to the Riviera may often wonder how so small a creature can make so great a noise. We have heard them in the hills about Alassio giving forth a sound as if some large steam-engine were hard at work in the vicinity.

Frogs that live on trees may, in spite of their adhesive finger tips, sometimes fall; and it would be a great gain to such creatures if they possessed anything like a helpful parachute, as the so-called "flying" squirrels and "flying" opossums—animals which have such a help—in an extension of the skin of the flanks.

Bats are flying beasts now, and pterodactyls were flying reptiles in former ages. Whether any batrachian can in any sense be said to fly, we will not venture to affirm, but there is certainly one tree-frog which seems as if its feet might at least serve as a parachute. Mr. Alfred R. Wallace, in his travels in the Malay Archipelago, encountered in Borneo a creature which he declares to be "the first instance known of a flying frog." Of this animal he gives the following account: "One of the most curious and interesting creatures which I met with in Borneo was a large tree frog which was brought me by one of the Chinese workmen. He assured me that he had seen it come down in a slanting direction, from a high tree, as if it flew. On examining it I found the toes very long and fully webbed to their extremity, so that when expanded they offered a surface much larger than the body. The fore legs were also bordered by a membrane, and the body was capable of considerable inflation. The body was about four inches long, while the webs of all the feet covered a space of about twelve square inches. As the creature was a tree frog, it is difficult to imagine that this immense membrane of the toes can be for the purpose of swimming only, and the account of the Chinamen that it flew down from the tree becomes more credible."

There are species of frogs and toads, some of them not otherwise different from their fellows, the young of which do not form gills, but develop some other breathing organs in their place. Thus, there is a kind of tree frog, belonging to the genus *Hylodes*, which has the habit of laying its eggs singly in the axils of leaves, and the only water they can obtain is the drop or two which may from time to time be retained there. The young is, most strange to say, provided with a special breathing organ in its tail. Another frog, with yet another abnormality, has been recently discovered by Mr. Guppy in the Solomon Islands. He tells us: " During a descent from one of the peaks of Faro Island, I stopped at a stream some 400 feet above the sea, where my native boys collected from the moist crevices of the rocks close to the water a number of transparent, gelatinous balls, rather smaller than a marble. Each of these balls contained a young frog, about four lines in length. On my rupturing the ball, the tiny frog took a marvellous leap into existence, and disappeared before I could catch it."

Thus these frogs are never tadpoles, nor was anything

THE BULLFROG 105

in the shape of gills to be detected beside the neck, nor yet any tail. There were, however, certain folds on each side of the body which may turn out to be peculiar temporary breathing organs, like the respiratory tail of the *Hylodes* before mentioned.

Another American tree frog, named *Nototrema*, has a

FIG. 27.

THE PIPA.

curious pouch which extends in the female over the whole of the back and opens posteriorly. Into this opening the eggs are introduced as soon as laid, and the young undergo their process of development in this large cutaneous maternal sac. Of course they have no opportunity of living the life of tadpoles. Neither have the young of the well-known kind next to be described.

This latter kind is the great South American toad called the *Pipa*, whose mode of reproduction was at first greatly, though very naturally, misunderstood. The skin of the female's back, as the time for egg-laying approaches, thickens greatly and becomes of quite a soft and loose texture. The male, as soon as the eggs are laid, takes them up and imbeds them in this thick, soft skin, which closes over them. Each egg so enclosed then undergoes the process of development to the end, so that instead of coming forth as a tadpole, each young one comes forth as a very small but fully formed toad. Here again the young never develop gills. When the young have gone forth, a little pit or depression marks the spot where each egg was developed, and as many as 120 of these pits have been counted on the back of a single female.

But a yet more singular mode of development takes place in another American frog which comes from Chili, and is known as Darwin's *Rhinoderma*.

Here nothing special is to be seen in the female, but in the male a large sac or pouch is present and extends beneath the skin under the whole surface of the belly and lower jaw. No external opening into it is to be found until we open the frog's mouth, and then we find two apertures, which lead directly into it, placed on the floor of the mouth, one on either side of the tongue. There are many animals, the males of which will eat their own offspring if they get the chance. They, however, perform the act for their own pleasure or profit. Not so Darwin's *Rhinoderma*. He takes his wife's eggs, indeed, into his mouth, but it is for *their* good, not his. He does not swallow them into his stomach, but passes them through the apertures on either side of his tongue into his great ventral pouch. There they develop and become lively young frogs, to the questionable comfort of their exemplary sire. When sufficiently developed, they make their way up into their father's mouth, and, from that gaping aperture, out into the wide world.

All frogs and toads are very much alike, the greatest differences depending on the tongue. Except one American species, the *Pipa*, one African, and one Australian, which three have no tongue at all, all the frogs and toads have a tongue, which, unlike our own, is fixed in front and free behind. A certain Mexican species forms a single exception, and is like ourselves in the mode of its tongue's attachment to the mouth.

All frogs and toads form together a very natural order of the class Batrachia, and this order of theirs is the order of "the tailless ones"—the *Anoura*, or, as they are sometimes called, the *Ecaudata*. Before, however, speaking of other batrachian orders and comparing the *Anoura* therewith, it may be interesting to our readers to be told of one or two exceptional points in the structure of frogs and toads generally.

The number of separate bones, or vertebræ, which makes up the backbone, vary in different animals; but none have so few as the frogs and toads. They have but nine at the most, and many have, as the *Pipa*, but seven. A long styliform bone is attached posteriorly to their vertebræ. The *Anoura* have also a noteworthy peculiarity in the foot. Our own feet are formed of seven short bones joined together in a cluster, and to these the bones of the toes are attached in front. The bones which are thus short and clustered in us, are short and clustered in all other animals save the frogs and toads and one or two animals (lemurs) which are often classed with monkeys. In these animals (frogs, toads, and certain lemurs) two of the bones which elsewhere are thus short, are lengthened out so as to form another segment to the hind limb. It is of course quite impossible to suppose that animals as diverse as frogs and those monkey-like animals the lemurs, could have gained this similarity of parts by inheritance. It is obviously a case of " the independent origin of similar structures."

A very instructive change takes place in the development of the frog in those structures which answer to our own bone of the tongue. This bone in us consists of a median central portion and two pairs of processes called the greater (or hinder), and the lesser (or anterior) horns. The young tadpole has, as fishes have, a series of arches on either side of the throat supporting its gills. As the animal develops these arches grow smaller and smaller, till in the adult frog there comes to be a tongue bone, with two pairs of processes or "horns" as in ourselves. It is the hinder pair of "horns" of the frog which are formed from its gill arches, and thence we learn that in our own pair of hinder, or greater, "tongue-bone horns" we have what answers to the gill arches of tadpoles and of fishes. Such is the case, because the gill arches of fishes answer to the gill arches of the tadpole.

Having now passed in review the more interesting forms of frogs and toads—the order *Anoura*—it is time to inquire what are the creatures which form the second order of the Batrachia ? As they all have long tails the name of their order is *Urodela*, or, as it is sometimes called, *Caudata*. Some such creatures are to be found in most ponds in England and are familiar animals to every schoolboy, and are known as efts or newts. The whole world contains a hundred and one different kinds of them, but of these no less than fifty-five species are found in North America, while Australia and tropical Africa have neither of them a single species. Only two are found in the Indian region, and but nine, or at most ten, in tro-

## THE BULLFROG

pical America, while the rest inhabit the northern portion of the Old World. Thus the order of newts is an order almost entirely confined to the northern hemisphere, whence a few struggle southward along the eastern Asiatic mountains or the Andes. China and Japan being the part of Asia nearest North America, are, as might be expected, rather rich in the number of their species. But it is not merely or mainly by the numbers of species that it contains that North America is thus

FIG. 28.

THE AMPHIUMA.

distinguished. The singular interest of some of the forms peculiar to it is even yet more striking. The absolutely largest species is, however, found in Japan, where it sometimes attains the length of six feet. A closely allied species is found in northern China, and during the tertiary period one also inhabited Europe. Its remains were discovered in the early days of geological science, and were taken to be the skeleton of a child, a victim of Noah's deluge. A much smaller representation of the Japanese giant eft is found in all the tributaries of the Mississippi and the streams of Louisiana, as well as in North Carolina, Ohio, and Pennsylvania, and is, at least in some places, known by the curious name of "hell-bender." A very singular form (Fig. 28), the *Amphiuma*, is also to be met with in the Southern United States. It has much the aspect of an eel, but with two pairs of very minute limbs, situated far apart, each with but three or even but two toes. This creature is, or was, called by the negroes "Congo snake," and quite erroneously regarded as very venomous.

The next noteworthy form is a European one. It is the *Proteus*—a small, entirely aquatic animal found in the subterranean caverns of Carniola and Istria, in southern

FIG. 29.

THE PROTEUS.

Austria. It is very elongated in form, with small and slender limbs and with as few toes, as the "Congo snake." Passing as it does the whole of its life in perpetual darkness, it is not surprising to find that it is blind, like so many inhabitants of American caverns. It is also colourless, except its gills, which project externally as a bright red tuft on either side of the hinder part of the head. These are constantly present, as this animal retains them during the whole of life, and is thus, so far, like a persistent young tadpole. The same thing occurs in two other species which inhabit the United States.

The first of these is more exceptional, in that it has no hind limbs at all, and only a very small pair of fore limbs. This is the *Siren*, specimens of which have come to the British Museum from various parts of the United States, including South Carolina, Georgia, and Texas. The other American form with persistent gills is named *Menobranchus*. It is shorter in body, with two pairs of fairly developed limbs. It is a more northern animal, being found in Canada as well as the United States. There is a genus of efts, distinguished both from the number and considerable size of the species which compose it. It is named *Amblystoma* and contains seventeen different species, all of which are North American

FIG. 30.

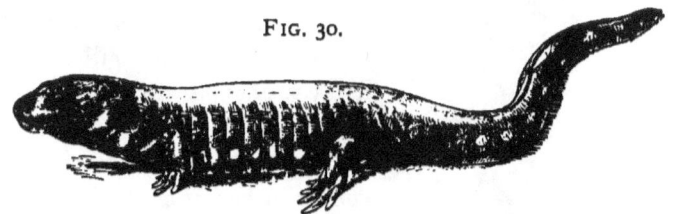

THE AMBLYSTOMA.

(ranging from the northern Rocky Mountains and Vancouver's Island to Mexico), with one exception—a single species which, strange to say, has found its way to the mountains of Siam! All these species when full grown are quite destitute of gills, though almost all efts (like almost all frogs) pass through a stage of existence in which they do have them.

There is a sort of eft in Mexico known as the *Axolotl*, which possesses large external gills. It seems a fully adult creature and breeds freely. Long ago, however, the great Cuvier considered that it probably was only an immature animal.

A little less than twenty years ago, there were a number of axolotls living in the Jardin des Plantes at Paris, under the care of a very intelligent keeper. One day to his astonishment he missed one of his axolotls, but found in its place a very different-looking eft, and one without any gills. A little later the same thing happened in the case of several of them, and thus it became known that the axolotl is but a big and precocious baby, ready to change rapidly, under certain conditions, into the form of an amblystoma. This change is a very remarkable one, because it by no means consists merely in the loss of the gills, but involves changes in the bones of the skull, the number and arrangement of the teeth, as well as other important structural transformations. The change was the more singular because, although the unchanged axolotls continued to breed freely in their immature condition without any care or trouble to their keeper, none of the transformed ones could be induced by any effort of his to do so. It seemed as if they had, on obtaining maturity, discarded all conjugal family feelings as mere follies of youth.

FIG. 31.

THE AXOLOTL.

This curious change in the axolotl, and its long persistence in breeding in a condition which, as far as form and structure go, must be regarded as an immature one, suggests a question whether the siren, the proteus, and the menobranchus may not also be overgrown babies, which have now ceased altogether to assume the form which once was their mature one. Individuals of the commonest English species occasionally preserve some of the external characters of immaturity, in spite of having attained reproductive capability. The Alpine eft also very often breeds before attaining the form which that species normally exhibits.

The true salamander—a very handsome black and yellow animal—is found from Holland to Spain, Algiers, and Syria. Like the pipa toad, it brings forth its young in the adult condition, they being born without gills. One result of the maturity they attain thus early is that one unborn brother sometimes devours another. Before birth they for a time have gills, and gills of relatively large size. Some curious experiments were tried on the dark Alpine variety of this animal by a Miss Von Chauvin. She took from the oviduct of a mother salamander some of its unborn young in that early stage of existence when they have large external gills. These specimens she placed in water, where the first effect of the disproportionately large size of their gills was that they were cast off. Thereupon, new and much smaller gills appeared in their place, and lasted a long time—fourteen weeks in one instance.

The curious and noteworthy point in this experiment is the fact that after the original gills (which were unadapted for free external life) had perished, new and suitable gills became developed, and this not in a struggle for existence against rivals, but directly and spontaneously from the innate nature of the animal.

The order of efts and the order of frogs include all the familiar forms of batrachian life. The next order contains only a few curious forms, which are confined to the warmer regions of the globe. The order which contains them is named *Ophiomorpha* (from their snake-like, legless condition), or sometimes *Apoda*. They are so strangely unlike the creatures hitherto described that no one would at first suppose they could have any affinity to frogs.

They are creatures which are entirely destitute of limbs and have much the appearance of earthworms. They are long and slender, and have a soft and naked skin marked by transverse, grooved rings. There are about thirty species, arranged in eleven genera. North

THE CŒCILIA.

America, Australia, and the northern part of the Old World, including all Europe, are entirely destitute of them. Their headquarters lie in tropical America, where one-and-twenty species have their home; four hail from tropical Africa, and a like number from the Indian region.

It is not surprising that these animals were classed with snakes by the earlier naturalists, and even by Cuvier. In spite of their small head, their being utterly deprived of limbs, and their elongated form, they have after all one notable point in common with those large-headed, long-limbed, and short-bodied creatures, the frogs. This point in common is the absence of the tail, for their elongated figure is due to the drawing out of the body, not to the presence of a tail.

They are creatures which burrow beneath the soil (which habit increases their resemblance to earthworms) and feed on any small creatures they thus find, and also upon mould.

Non-scientific persons may perhaps ask, Why should these worm-like, limbless creatures be grouped in the class Batrachia with frogs and efts? It is an extremely natural question, but one which admits of a very easy and satisfactory reply, although we cannot venture here to inflict on our readers the strong dose of anatomy which would be necessary to set out with any fulness what that reply is. We may, however, call attention to one character, as follows: No beast or bird or reptile develops gills at any time of life, while all the batrachians do have them, and breathe by help of them at first, with the exception of a few species, the conditions of whose life render such aquatic breathing impossible for them. There are, indeed, as we have seen, very exceptional non-gilled forms, but these plainly and evidently are the brothers of those other kinds which do have gills, which form the overwhelming majority of the whole class.

The frogs and toads (*Anoura*), the efts (*Urodela*), and the worm-like kinds (*Ophiomorpha*), comprise all the orders of existing batrachians. Another order, however, is known to us, but to obtain evidence of that order it is necessary to dig very deeply into the geological evidences of the past.

When we begin our search into the geological records of bygone ages, we do not find evidence of anything startling or new in the tertiary or even in the upper secondary rocks. Frogs and efts have indeed been found in the tertiary strata, but they differ in no important way from their representatives of our own time.

When, however, we descend to the lias and trias and carboniferous rocks we come upon a great variety of extinct species of animals which have been hitherto regarded as being nearly allied to the three batrachian orders still existing. They constitute another—fourth—order, to which the term *Labyrinthodonta* has been applied for a reason which will be stated shortly. Thus our search into the past has brought us a rich and important gain. The labyrinthodonts were creatures with long tails and mostly two pairs of limbs, but their limbs were always of relatively small size. Some species were very large, even exceeding the great salamander of Japan in size. Some of them had large and formidable teeth in long elongated jaws like those of alligators, and the structure of their teeth is very noteworthy. They are conical in shape and marked superficially by slight vertical grooves. If we make a horizontal section of one of their teeth we shall then see that these surface grooves are the external indications of deep indentations of the substance of the tooth. All these indentations converge toward the central long axis of the tooth, but do not converge in straight lines, each indentation being elaborately inflected. Radiating from the central axis of the tooth we shall find in our section a corresponding number of grooves radiating outward from the tooth's central pulp cavity, these radiating folds passing between, and being inflected in undulations like, the converging grooves. Such a tooth is a beautiful object when examined with a good magnifying glass, and the markings thus produced by so many radiating and converging folds (undulating and alternating in regular order) are so complicated and labyrinthic, that labyrinthic tooth becomes an appropriate name for the members of this singular order. Its members were doubtless aquatic in habit, as are almost all existing batrachians; but to which group of the latter can these ancient monsters be considered to have affinity? It is quite impossible to affirm that they in any way tend to bridge over the chasm which separates the frogs from the efts. They appear, indeed, to have been almost equally removed from both. It is not impossible that they may find their nearest existing allies among the worm-like Ophiomorpha. There is a curious resemblance between the skulls of these two groups, which also agree in containing more bones than do those of the members of the other two orders. Moreover, some labyrinthodonts appear to have been entirely deprived of limbs. Nevertheless, their largely developed tail constitutes a marked distinction between them and the Ophiomorpha.

It is somewhat singular that in spite of their predominating aquatic habit all batrachians appear to inhabit fresh water only, and they are thus the only class of backboned animals which have no marine representatives. But as regards the labyrinthodonts, it is very probable that many new and strange forms will come to light, and we look forward with great interest to such a revelation with regard to these air-breathing animals of the carboniferous epoch, which, so far as we know, were about the earliest air-breathing vertebrates. We have noted already, in describing the turkey, how different forms of reptilian life preceded and represented the beast life of our own age—its whales, its bats, its cattle, and its beasts of prey. Reptiles represented them during the vast epoch which continued while the secondary rocks were being deposited. The number and variety of labyrinthodonts already found suggests the idea that a great wealth of batrachian life may have preceded and represented the reptile life of the secondary age. They may have represented its crocodiles, ichthyosauri, dinosaurs, &c., during the vast epoch which continued while the carboniferous rocks were being deposited.

This, however, is as yet a mere speculation. What is certain is that the batrachians stand (as before said) between reptiles and fishes. To the fishes they are manifestly allied through their possession of gills and the aquatic respiration they almost all practise during the earlier stages of their existence. There are also other anatomical points of resemblance, which it would be tedious here to describe. On the other side they exhibit a striking difference from all fishes and a close resemblance to reptiles and higher vertebrates in the structure of their limbs. In ourselves we have an upper and lower limb-segment to both arm and leg, we have a cluster of small bones in the root of both hand and foot, and we have fingers and toes (both called in zoölogy digits) proceeding therefrom. With varying degrees of defect, the same essential structure of limb exists in all reptiles and higher vertebrates, but no such structure exists in any kind of fish. Reptiles have, fishes have not, this "typical differentiation," as it is called, of the limbs. Thus, batrachians let us down gently, as it were, to the class of fishes, while retaining a firm grasp—with their typically differentiated limbs—on the class of reptiles. They do so in different degrees, there being but a temporary affinity to piscine respiration in the frogs, though a permanent one in the proteus and the siren.

While differing insignificantly in structure among themselves, we have seen how very widely frogs all differ from the other orders of their class—that is, from the efts (*Urodela*), the worm-like group (*Ophiomorpha*), and the primitive batrachian group (*Labyrinthodonta*).

## THE BULLFROG

The bullfrog, indeed, pertains to an order far more distinct from the other orders of its class than is man's order from the various orders which compose his own class, the class of mammals. It also belongs to an order which is singularly homogeneous, and yet to a class (Batrachia), which compared with that of birds is very heterogeneous.

It is an animal which differs from every member of the higher classes of vertebrates in that it comes into the world with a structure and with habits which contrast most forcibly with its structure and habits when adult. In fact, it undergoes a metamorphosis!

If, then, we are asked, What is a bullfrog? we may reply: "It is a very large North American species of the genus *Rana*, a genus of an order of tailless, lung-breathing, gilled vertebrates, with fore limbs typically differentiated and undergoing a distinct metamorphosis, its order being one of those four which makes up the class Batrachia."

Such is our reply to the question as to what the frog is; but we may further ask how did it come to be—what was the origin of frogs? We may also, on the principle of evolution, and seeing how very ancient the frog's class is, be asked what forms may be supposed to have sprung from it? Of what existing creatures, which are not Batrachia, can batrachians be supposed to have been the ancestors?

In the early days of the promulgation of the theory of evolution nothing seemed easier than to answer such questions. Genealogical trees of animal life were set up by very many naturalists—most conspicuously of all by Prof. Haeckel of Jena—with eagerness. Soon, however, they were found to need pruning, then "lopping and topping," and finally not a few have we seen cut down or torn up by the roots. Some of our own modest shrubs we have come to believe merit the same fate, though we have not to answer for much such arboriculture, on account of our having from the first believed in and called attention to the "independent origin of different structures."*

It was at one time believed rather widely that the eft group sprung, through the labyrinthodonts, from certain air-breathing fishes, and, in turn, gave rise to frogs on the one hand and to beasts and reptiles on the other. This may be all very true, but as yet we must regard it as a mere speculation, without any sufficient evidence of its truth to hinder us keeping quite "an open mind" about it. We have already seen, with respect to the opossum's order and class, how two different and contradictory hypotheses may be suggested by one set of facts. As to the above-mentioned belief we do not think there can be any reasonable doubt about its first article —that batrachians sprang from fishes. What kind of fishes those were, however, we do not know. It is also most probable that frogs did also spring from eft-like creatures, but the utter absence to his time of any discovered links between the two is very remarkable.

But the once asserted direct affinity between batrachians and beasts we do not at all believe in, or in any affinity between existing batrachians and reptiles, though very probably the first reptiles sprang directly from some ancestor or collateral relative of the labyrinthodonts. The resemblance of frogs to tortoises and terrapins has often been remarked, and it is remarkable. It is clearly, however, but an instance of the independent origin of similar structures, and a direct descent of tortoises from frogs is quite incredible. It is none the less interesting

* In our "Genesis of Species," 1870.

to note that in the mud-tortoises we have the bony plates of the shell greatly reduced and surrounded by soft skin, while in two kinds of frogs the skin of the back becomes furnished with bony plates which are complete representatives of those of the tortoise, though much smaller.

Again, the turtle is exceptional among reptiles for such an extension of some of the skull bones as to give the brain case a deceptive appearance as to size. The very same thing is also found in two members of the frog's order. How cautious it is necessary to be in attributing such similarities to special inheritance has been strikingly shown by the discovery in an African animal belonging to the rat's order (Rodentia) of the very same kind of extension of the bones of the skull. Building upon such resemblances it might be supposed that frogs were the parents of tortoises, efts of lizards, and the worm-like ophiomorpha of snakes. But here we hope enough has been said to show that such a view is utterly false, as also to impress on our readers a wholesome caution as to wild speculation and hasty generalisation in matters zoölogical.

While keeping our minds free from prejudice and ready to receive all and any truth which may be demonstrable, we must be scientifically exacting in our demand for evidence with respect to all hypotheses put before us.

V

## THE RATTLESNAKE

AMERICA has the privilege of possessing a variety of interesting animals found nowhere else in the whole earth, such as many kinds of apes, opossums, and numerous birds; above all, its lovely humming birds, the exclusive possession of which every other quarter of the world may well envy. But it also possesses exclusively some creatures, the presence of which will not excite envy in other geographical regions. Among these are the rattlesnakes, which, in their various varieties, range from Southern Canada down to Patagonia. But, deadly and dreadful as they are, they are creatures nevertheless full of interest for persons who love the study of Nature, her works and ways among living things. The rattlesnake has an interest for a variety of reasons—(1) on its own account, (2) as one of a small group of poisonous serpents, which includes also forms which are not rattlesnakes, (3) on account of the relation in which it stands to all other snakes, and (4) as being a snake at all; for every snake is a very remarkable animal, and probably many of my readers do not know what a snake really is. If so, then they necessarily must have but a very imperfect comprehension of the deadly American reptile. To be able in a satisfactory manner to answer the question, "What is a rattlesnake?" we must know something definite as to what any snake is, as compared with all creatures which are

Fig. 33.

THE COMMON RATTLESNAKE.

not snakes; as to what snakes are, considered as a group in themselves; and as to what are the exact resemblances and differences which rattlesnakes, and every particular kind of rattlesnake, bear to all other kinds of serpents.

Of rattlesnakes there are at least a dozen, probably fifteen, different species, though there are a good many varieties; a fact which makes them difficult to define. The kind most common east of the Mississippi is popularly known as the "banded rattlesnake," and ranges at least from Maine to Texas. At one time it was very common in Eastern Massachusetts, where it is now, happily, very rare indeed, and only common in thinly inhabited districts of more Southern and Western States. It varies a good deal in colour, and may be mainly brownish, yellowish, or blackish, while a series of dark spots, frequently edged with yellow and of very variable shape, run along the back and sides. The head is very large, much flattened and triangular in shape, the exterior angle being rounded. One very noticeable feature is a deep pit which is placed between the eye and the nostrils on either side of the head. The use of this structure remains unknown. The snake often attains a large size, that is, five feet in length. It feeds on rabbits, squirrels, rats, &c., and is for the most part slow and sluggish, waiting quietly until some suitable prey approaches it. The notion formerly entertained that the rattlesnake can charm or fascinate other creatures is a mere superstition, now quite exploded. But its sluggishness makes it dangerous, as it may be unknowingly stepped upon. Yet it never attacks spontaneously, or pursues a retreating enemy.

The structure from which the animal takes its name— the "rattle"—consists mainly of three or more solid horny rings placed at the end of the tail. These rings themselves are mere modifications of the general skin of the body, but the "rattle" has a more solid foundation. The real tail of birds (as we saw when considering the turkey) is made up of a short fleshy structure supported by a special modification of the terminal segments of the backbone. The same is the case with respect to the rattle of this serpent. Its three terminal segments (or vertebræ) become united together into one solid whole, and also become enlarged in size and specially modified in form, being swollen at the hinder end. This bony structure is covered with a special development of the soft deeper skin from which all the outer skin and scales of the body are formed, the soft structure being so subdivided by grooves as to form three segments, which themselves become coated with three corresponding dense layers of outer skin, or, as it is technically called "epidermis," thus forming three horny rings. These constitute all the rattle there is in young snakes which have not yet shed their skin. Snakes and men shed their skin differently. In us the outer skin is thrown off in very minute separate portions, so that the process is not ordinarily, noticed. In snakes all the skin is shed at once as one continuous whole—even the skin of the eyeballs being shed with the rest, and thus snakes get a little blind during the process of its detachment. When the first moult in rattlesnakes draws near, fresh skin is formed beneath the old covering of the hinder end of the tail. When the moult actually takes place, the old covering of the tail end is not cast off (being held by the swollen end of the bone before noted), but remains as a loose appendage, thus becoming the first formed joint of the future perfect rattle. The rattle, in fact, grows perfect by the accumulation of rings in this manner, one being thus made loose and yet retained, at each succeeding moult, while more than one moult takes place in each year. Thus the rattle ultimately consists of a number of dry, hard, more or less loose, horny rings. The older of these wear away in time and are lost, but a snake may have as many as twenty-one rattling rings.

It is the shaking of these rings by a violent and rapid wagging of the end of the tail that produces the noted sound—a sound which may be compared to the rattling of peas quickly shaken in a paper bag But this habit of shaking rapidly the end of the tail is by no means peculiar to the rattlesnake. It occurs in many other species of serpents, both venomous and harmless ones. It is probably a natural and spontaneous result of emotional excitement, like the wagging of a dog's tail. Any nervous excitement tends to produce some bodily movement, and naturally results in the motion of any part most easily moved—as the end of the tail, whenever a due supply of muscle exists to produce it. The meaning or use of the rattle is a problem still awaiting solution. It has been supposed to be useful in paralysing animals, through terror excited by the sound; in arousing curiosity and so bringing animals within its reach; as enabling the sexes to find each other; or by saving the snake from attack when its power of offence temporarily has been exhausted. But no sufficient evidence known to us lends adequate support to any of these ingenious speculations.

Among the various species of rattlesnake is the kind called the "horned rattler," on account of a pair of horny prominences it possesses, one above each eye. It is found in California and Mexico.

The deadly bite of the rattlesnake is effected by means of a very ingenious and simple mechanism. It is a popular error to suppose that the rapidly vibrating cleft

## THE RATTLESNAKE

tongue of the creature, so often protruded from the front of its muzzle, is its "sting"; the rattlesnake poisons by biting, and the only practical sting it possesses consists of a pair of peculiarly modified teeth. The lower jaw is furnished on either side with a series of small, simple-pointed teeth, and two series of small, simple-pointed teeth traverse the palate from before backward. The outer margin of the upper jaw, however, has nothing of the kind, but is furnished instead on either side with one large, powerful curved and very pointed tooth, which is the "poison fang." This poison fang is very deeply grooved in front. It is, indeed, grooved so deeply that the two margins of the groove quite join in front, save at its upper and lower ends; the groove is thus practically converted into a canal which traverses the substance of the tooth. Into the upper unclosed end of the groove a small tube passes, and this conveys the poison from the gland which secretes it, into the cavity of the tooth. It then passes down the canal and escapes from the small unclosed end of the groove which opens near the point of the tooth. The poison gland is placed on either side of the upper jaw (extending backward beyond the eye), and the poison itself is but a form of saliva. Its deadly effect almost every one knows. Even if an adult man escapes with his life he must suffer from prolonged illness and often from the loss of a limb. When the rattlesnake is at rest, the poison fangs lie back against the roof of the mouth, but when excited, as he opens his mouth the fangs become erected by a peculiar mechanism which cannot be here described, as its description would involve so many technical anatomical details. Suffice it to say, that when, being erected, the snake strikes, his poison fangs bury themselves in the flesh of his victim, while simultaneously the poison is ejected down the canal which traverses each of them. The feeling of anger also doubtless sets the poison glands secreting, just as the sight of good food will make a hungry man's mouth water—*i.e.*, will set similar glands secreting in the man's mouth. The rattlesnake strikes its prey to kill it, and, having struck, waits quietly till it dies. Then it begins to devour it at leisure, not again using its fangs, but only the small teeth before mentioned. It always devours its prey entire, and can swallow an animal much thicker than its own body. In fact, the creature does not so much swallow its prey as slowly drag itself over the creature it devours, being enabled so to do by the elasticity of its skin, and by the extraordinarily loose condition of the teeth-bearing bones of its jaws. Thus the two halves of the lower jaw and the several pieces which compose the upper jaw can be stretched far apart and separately moved, so that, while the dead victim is securely held by some of them, others can be moved and implanted further on, and thus, by degrees, its body is drawn within the gullet of the snake. Even when it has passed into the stomach the form of the prey may be visible for some time, but digestion takes place very quickly. The rattlesnake has no rudiment of a limb, and its movements are effected by nothing but its backbone and ribs, with the aid of the muscles thereto annexed (which are very numerous and complex), and that of the large transverse scales which clothe the abdomen. The ribs are very movable, and their lower ends are connected to the inside of these scales. Thus the snake's motion may in part be compared with that of a centipede, the successive opposite pairs of movable ribs practically serving as so many pairs of feet. It is thus, with the aid of the ribs and scales, that the rattlesnake progresses by taking advantage of the various irregularities of the surface over which it moves. On a perfectly smooth surface it can make no advance at all. It is common enough to see serpents represented in figures as bending their body in a series of vertical folds. This is another mistake. A snake's body can be bent only from side to side.

The rattlesnakes form part of a small group of serpents some of which have no rattle. They all agree, however, in being very poisonous and in having the curious pit already described as placed between the nose and the eye. The whole group (rattlesnakes included) are therefore spoken of as "pit vipers." Some of the pit vipers which are not rattlesnakes are found in the Old World, while others are (like rattlesnakes) American reptiles. Among the latter are the copperhead and water-moccasin of the Carolinas and Texas (so dreaded by workers in rice plantations), and the even more ferocious *fer-de-lance* of the West Indies, which attacks without warning, and is said to have been the main cause of death among the labourers in sugar plantations, wherein it finds shelter and often multiplies prodigiously. Lastly may be mentioned Bushmaster (*Lachesis*) of tropical America, which seems to be the largest poisonous land snake known, as it is said to be sometimes fourteen feet long. In the Old World a dozen species of pit vipers are found, in India one species ascending as high as 10,000 feet above the sea in the Himalayan mountains.

The whole group of pit vipers, including the rattlesnakes, forms but one subordinate section of the great order of serpents, which order contains as many as 1500 species at the least. These are generally divided into the poisonous and non-poisonous snakes; but such a division is not a natural one, for some poisonous snakes are much more closely related to the non-poisonous kinds than they are to the other venomous forms. The non-poisonous kinds are far the more numerous, but nevertheless some 20,000 human beings are killed every year in India alone by venomous serpents.

No distinct chemical principle has yet been detected in the poisonous saliva, but such a thing there must be in the different kinds of snakes, seeing that after a fatal bite from a rattlesnake or viper, the victim's blood will not coagulate, while after a fatal bite from a cobra it will still do so. It has also been ascertained that if the blood of a bitten animal be injected into a healthy one, the latter will be poisoned just as if it had been bitten itself, although its flesh may be eaten with impunity. It is a mistake, however, to suppose that snake's poison can have no effect unless actually mixed with the blood. It will act after being absorbed through such delicate skin as that which lines our lips, though its action is then less powerful.

The effect of snake bite depends partly on the condition of the snake, partly on that of the person bitten. If the snake has bitten shortly before, or if it is not in a vigorous state, its effect will be more or less diminished; while it is increased if the person bitten is weakly, very nervous, or a child. As to nervousness, some persons are said to have died merely of fright. Of course, something depends on the part of the body bitten, a bite being especially fatal if the fangs actually penetrate a large blood-vessel. The bite of a rattlesnake has been known to produce almost instant death. No effectual antidote has as yet been discovered. Ammonia and permanganate of potassium are ineffectual, although a solution of the latter will take away the poisonous effects of the snake's secretion if it be mixed with it. Immediate amputation of the bitten toe or finger is the best course, as the delay of a few seconds will suffice to convey the poison into the patient's circulation. If amputation cannot be thus performed, a very tight ligature, with sucking and cauterising the wound and the administration of stimulants internally, are recommended as the best treatment.

The whole order of serpents may, for our present purpose, be divided conveniently into four great groups: A, viperine snakes; B, colubrine snakes, so called from the name coluber, applied to a large genus of these snakes in both hemispheres, and originally instituted by Linnæus; C, boa-like snakes, and, D, worm-like snakes. Having selected the rattlesnake as our type, we will begin with the first section, A.

This consists of two groups, the true vipers, such as the common English viper, and the pit vipers. The latter we have already considered. The true vipers are as exclusively confined to the Old World as are the rattlesnakes to America. Among the latter we have noted one as being termed the horned rattler. The same epithet is most justly applied to various true vipers, some of which have horns over the eyes, while others have two such structures on the nose. Among the former is the famed *Cerastes* of Africa, which imagination has connected with Cleopatra. A far more magnificent creature is the rhinoceros viper (Fig. 34). It is a very deadly animal, which may be more than six feet long and beautifully coloured, with a pair of long horns upstanding from between its nostrils. There are at least twenty kinds of true vipers, and such are the only poisonous reptiles of Europe. Russell's viper, known as the "tic-polonga," or Daboia, is one of the most deadly snakes of India, while it is so sluggish that very often it will not move out of a man's way. Another most dangerous viper, though a small one, is the *Echis*, which is found in desert regions from Morocco to the middle of Hindostan.

Though little more than two feet long, it is not only fierce, but positively aggressive, while its poison is so active that a fowl dies about two minutes after being bitten. Two or three species of vipers of slender form and adapted for tree life are also found in Africa.

FIG. 34.

THE RHINOCEROS VIPER.

The colubrine section of serpents (B) contains the great majority of the species which compose the order, including the commonest harmless snakes of Europe and North America, such as the black snake, the corn snake, the milk snake, and the chicken snake. The genus to which the common ringed snake of England belongs, ranges throughout the temperate region of the northern hemisphere. Other kinds there are which are very deadly, such as the cobra and various harlequin snakes. But, innocent or harmless, all these snakes, instead of having the great, conspicuous poison fangs of the rattlers and vipers, have a series of teeth extending along either side of both the jaws, as well as two rows of teeth in the palate—six rows in all. They feed, however, as does the rattlesnake, but with the disadvantage of having to overcome and engulf a prey still living. The common English snake will eat mice, lizards, or young birds, but its favourite delicacy is the common frog. When pursued by a snake, the frog seems to be half paralysed with fear, leaping less and less powerfully as the snake comes upon it. It is usually seized by the hind leg, but should it be taken by the middle of the body the snake invariably turns it till, by dexterous movements of its jaws, the frog's head comes to be directed toward the throat of the snake, and then it is swallowed head foremost. In menageries two or more snakes will often seize upon the same frog, when each one begins to swallow it from the point to which it has attached itself. Soon, however, the jaws of the rival snakes come in contact, and then follows a decisive struggle. On one such occasion Mr. Bell, the late well-known English naturalist, observed such a contest. He tells us (" British Reptiles," p. 51), that " On placing a large frog in a box in which were several snakes, one of the latter instantly seized it by one of its hinder legs, and immediately afterward another of the snakes took forcible possession of the fore leg of the opposite side. Each continued its inroads upon the poor frog's limbs and body, until at length the upper jaws of the snakes met, and one of them in the course of its progress slightly bit the jaws of the other. After one or two such accidents the more powerful of the snakes commenced shaking the other, which still had hold of the frog, with great violence against the sides of the box. After a few moment's rest the other returned to the attack, and at length the one which had last seized the frog, having a less firm hold, was shaken off, and the victor swallowed the prey." The frog is generally alive, not only during the process of deglutition, but even after it has passed into the stomach. Mr. Bell once saw a very small one which had been swallowed by a large snake in my possession hop again out of the mouth of the latter, which happened to gape, as they frequently do immediately after taking food. On another occasion he heard a frog distinctly utter its peculiar cry several minutes after it had been swallowed by the snake.

This reptile is easily tamed, and will learn to distinguish those who feed and caress it. It will sometimes nestle spontaneously within the folds of its master's garments and hiss at a stranger who would meddle with it.

One hundred and twelve species of harmless colubrine snakes inhabit India. Among the most attractive are the delicate tree snakes (*Dendrophis*), which very rarely descend to the ground, finding food enough among the birds and those kinds of frogs and lizards which also dwell in trees.

The venomous colubrines differ from the others in having some or other of the teeth of their upper jaws grooved. Among them are other tree snakes (*Dipsas*) found in Africa, South Asia and North Australia, and singularly beautiful in coloration, and the yet more slender arboreal whip snakes, with very long pointed heads (*Dryophis*), which are nocturnal in habit, and feed mainly on birds. One kind is very handsome, being black with a multitude of golden spots. A closely allied Indian form is very singular in its resemblance to a curious African kind, which might be called the "gullet-toothed snake" for the following reason : If its jaws be opened they will be found to contain but a few exceedingly minute teeth, but if the finger be passed down its throat, then a series of bony prominences projecting down from the under surface of the backbone and the upper wall of the gullet. Each of these bony prominences is capped with enamel, and acts as a true tooth. These snakes feed on eggs, and are so notorious that they are known in Cape Colony as the "egg eaters." Now, a snake's mouth is not bordered by any fleshy lips, and if this snake were to crack an egg in its mouth, most of the contents would run out and be lost. Accordingly it swallows each hen's egg whole, and then, when safely within the gullet, it squeezes and breaks the egg against the curious teeth of its backbone. Thus the nutritious contents of the egg is secured, and the waste, otherwise inevitable, entirely avoided.

The above-mentioned harlequin snakes of America (*Elaps*) are very handsome reptiles, their bodies being encircled with black, red, and yellow rings, as are also some American snakes which are not venomous. They are not large, rarely exceeding three feet in length, while both their mouths and poison fangs are small. Added to this, they only bite under great provocation, so that they should be little dreaded.

Forms allied to these snakes and those next to be described constitute the bulk of the serpents of Australia, that region of the world being distinguished from all the others by having the decided majority of its snakes venomous.

A small serpent of south-eastern Asia, called *Adeniophis*, is very remarkable for the exceptional size of its poison glands, which extend back for fully one-third of the reptile's entire length, so as to push the heart back much behind its usual place. We have as yet no information as to what may be the special utility to the animal of such an extraordinary supply of venom.

No snakes, not even the rattlesnakes, are more

Fig. 35.

THE INDIAN COBRA.

dreaded, and with reason, than are the cobras, also called "hooded" or "spectacled" snakes. As the rattlesnake warns the ear by its significant rattle, so the cobra warns the eye by the mode in which it expands the hood when irritated. The "hood" is a lateral expansion of the body just behind the head. This flattened expansion is produced by the sudden elevation of the ribs there situated, which stretch out the skin of either side as they rise. On the back of this hood there is a peculiar mark, roughly, like a pair of spectacles, hence the second trivial name of the creature.

It is the cobra which is chosen by the so-called "snake charmers" of both Egypt and India for their performances. The Egyptian ones sometimes pretend to change the serpent into a rod, and according to a French naturalist (G. St. Hilaire) this appearance can be produced by giving a strong squeeze to the neck, so inducing a convulsive rigidity, from which it soon recovers. It need hardly be said that snake charmers always carefully extract the fangs of their snakes before playing with them. The danger of otherwise touching such animals was sadly illustrated a few years ago by the act of a keeper of the Zoölogical Gardens of London, who incautiously took hold of one in his hand, and was immediately bitten. Before any effective aid could be rendered, the unfortunate man was a corpse. The Indian cobra attains more than six feet in length, and is the most generally fatal of all Indian serpents, being so common and widespread. It is found from the shores of the Caspian and southern China to the end of the Indian Archipelago. A second Indian kind, the snake-eating snake, is far larger and fiercer than the first. It may be fourteen feet in length, and is said actually to pursue and attack men. Fortunately it is much less common than the smaller species, though its distribution is as widespread.

The African cobra ranges from Egypt to the Cape of Good Hope, and that it was well known in northern

Africa thousands of years ago, is shown by its constant appearance in Egyptian hieroglyphics. Two other African kinds are known as the "sheep-stinger" and the "spitting snake." The latter is especially bold and active, readily attacking any one who approaches near it. In confinement it is generally very savage, opening its mouth and erecting its fangs, from which poison often may be observed to drop, and sometimes even to be ejected forcibly by the pressure of the jaw muscles on the poison glands. It is this circumstance which has given the serpent its name. The last group of colubrine snakes to which reference will be made here, is a very singular one. The existence of the famed sea serpent has been much disputed, but that sea serpents, not so famed, really exist it is utterly impossible to deny. There are about fifty species of these marine reptiles, all highly poisonous, but not practically dangerous, as they never quit the water and swim away rapidly at the least alarm. Their main home is the Indian Ocean, extending thence toward Madagascar, down to the coast of Australia, and across the Pacific to the western coast of South America. They also advance northward to the shores of Japan. Like all other serpents, they are air-breathers, and to help them to rise quickly to the surface of the water and to swim with rapidity the end of the tail is flattened from side to side. In order to breathe the more easily and securely, their nostrils are placed at the very end of the muzzle, and are protected with valves to secure them from an unwelcome influx of water.

Unlike other snakes, they cast their skin in small pieces, and unlike most snakes, though not unlike vipers, they bring forth their young alive without laying eggs. Their progeny can, of course, swim as soon as they are born. Their eyes are not adapted to see well out of

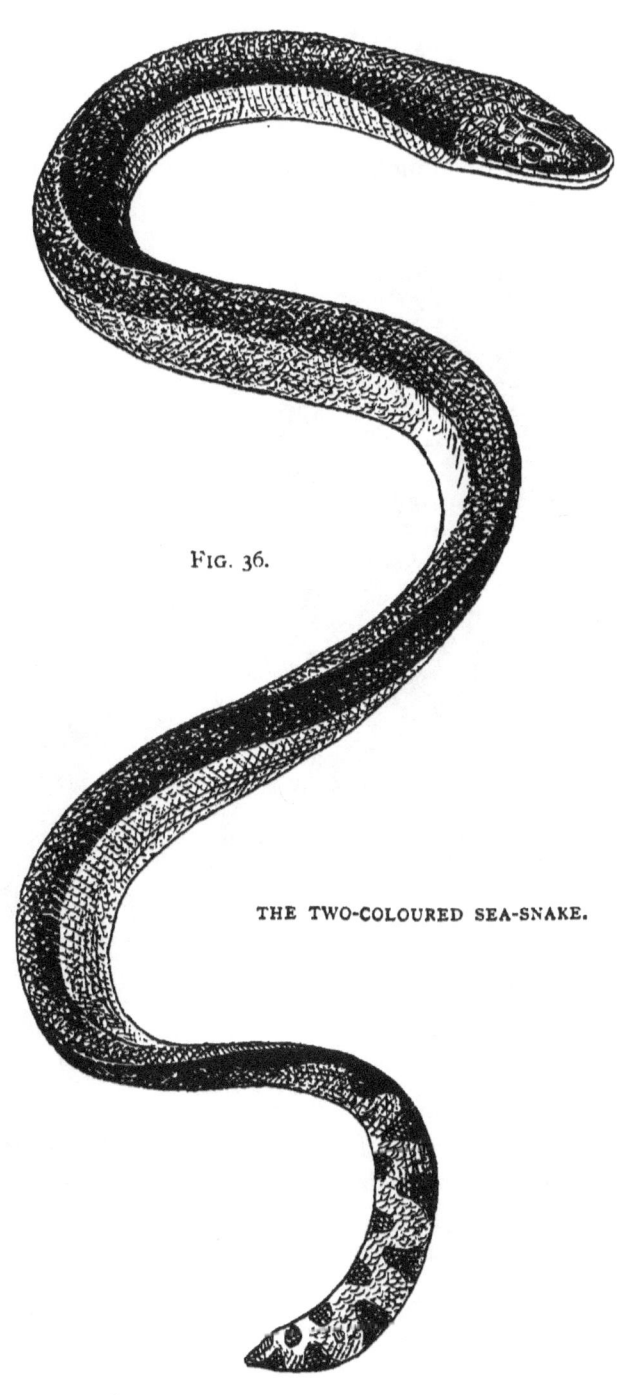

Fig. 36.

THE TWO-COLOURED SEA-SNAKE.

140    TYPES OF ANIMAL LIFE

water, and thus they cannot take good aim to bite. They cannot even live on land, nor has any one yet succeeded in retaining them alive for any considerable

Fig. 37.

THE BOA-CONSTRICTOR.

time in an aquarium. They feed on small fishes, which they paralyse with their poison and so have nothing to fear from their spines.

Snakes of the boa-like section (C) are few in number,

but are the giants of the serpent order. They consist of two groups, distinguished by no very striking difference: (1) the boas, and (2) the pythons. The former are confined exclusively to America and Australia and the tropical Pacific Islands, while the pythons are mostly from the Old World.

The kind so familiar by name, the boa constrictor, is found from the northern part of Central America to southern Brazil, and is very often seen in menageries. It is not a very large creature generally, not being much above seven feet in length, though it may attain fourteen. The boas and pythons both agree with and differ from the viperine snakes as regards their mode of feeding. They agree with them in that before proceeding to devour a prey they kill it. They differ from them in that they do not poison their prey, since they have no poison fangs or any grooved teeth. They kill their victims by crushing, and they perform the act with amazing rapidity.

We have often observed boas and pythons do this in captivity, and can affirm that the rabbits and duck introduced into their cages are entirely destitute of fear or apprehension and suffer nothing until they are seized, and then their sufferings are extremely brief. Such a serpent, if disposed to feed—to attain which disposition it often needs a fast of several weeks —will move slowly about till it brings its mouth opposite to the muzzle of the rabbit. Then in an instant its mouth is opened and the rabbit's head is seized, while simultaneously the voluminous folds of the powerful body are twined round it, and it is crushed immediately to death. The serpent does not at once uncoil its folds but continues for a time tightly to embrace its victim, so that reanimation becomes impossible. Then the monster slowly unwinds, takes the head of its prey in its mouth, and, by successively implanting, withdrawing and advancing its six rows of teeth (four in the upper jaw and two in the lower, as before described) gradually drags itself over the body of its victim till the latter is finally engulfed. The anaconda of tropical America is the largest serpent of the boa group. Its whole body is ornamented with dark round spots, and may attain a length of thirty feet. It is more or less aquatic in its habits, tenanting the banks of rivers and lakes, and lying in wait for animals which come to drink. It will kill and swallow a peccary or deer, the anterior parts of its body becoming, of course, enormously dilated during the process of deglutition. This process is greatly facilitated by a very copious secretion of saliva, with which the creature swallowed is abundantly lubricated. After a hearty meal it will, as before said, make a long fast. An anaconda in the London Zoölogical Gardens once remained more than three months without eating.

The pythons or rock snakes, as they are also called, are, as before said, mostly Old World forms. Three kinds are known in Africa and two in Asia, and some snakes which are similar, save that they are much smaller, are also found in Australia and New Guinea. Pythons may also attain a length of thirty feet, and will easily swallow a half-grown sheep. We have several times seen a python take three rabbits in rapid succession, or a rabbit and a pair of ducks. Pythons are fierce animals (save, of course, when in the torpid condition after food), but they are never spontaneously aggressive as regards man.

Most snakes lay eggs, but some of them, notably vipers and sea snakes, as before observed, hatch their eggs internally and so bring forth their young alive. Snakes' eggs have no hard shell like that of birds' eggs, but are enclosed in a leathery case, oblong in shape. The pythons actually incubate their eggs, as was first ascertained in the Jardin des Plantes at Paris. The female arranged her eggs in a conical heap and twined herself around them, her ever-watchful head surmounting the summit of the cone. The period of incubation will last two months, during which she will not feed, though she has been known to drink copiously if water were presented to her by her keepers. In spite of the generally cold-blooded condition of snakes (as of other reptiles), it has been well ascertained that the temperature of the mother is raised distinctly higher than that of the surrounding atmosphere during the process of incubation.

Various forms of small snakes, which here need not be separately noticed, lead us to our last section (D).

Its members indeed widely differ from all those I have hitherto described. Not only is the head small, but the small jaws are quite incapable of that mobility which is so extraordinary and so characteristic of snakes generally. Within their little mouths there are only a very few simple teeth, the lower jaws and the palate having none. Like the boas, they have rudiments of hinder limbs beneath the skin.

These snakes have the habits of earthworms, whence they are often termed "burrowing snakes," and many species are very much smaller than any ordinary earthworm. Their body is about the same size throughout, being clothed with small smooth scales in harmony with their burrowing habits. They have but a rudimentary eye. None of them are poisonous; their food consists of grubs, insects, and other small creatures. They are found all over the warmer regions of the world, especially in Asia and Australia. They are numerous in America, but extend to the shores of the Mediterranean and Japan.

We have now advanced considerably in reply to our question: "What is a rattlesnake"? We have seen what it is in itself and what is its position in the whole order of serpents. We now know that rattlesnakes form a portion of that subordinate division of viperine snakes which are known as pit vipers. We know that, poisonous as they are, they are not more so than some kinds of the great group of mostly harmless snakes, which goes by the name of colubrine. We have seen their relation to the other sections, and how they differ from the boa-like snakes and worm-like snakes. It only now remains to consider what a rattlesnake is in so far as it is a snake, and what are the relations which it thus bears to other reptiles.

Reptiles are creatures which form a class by themselves, but one which had very special relations with the state of this planet at epochs so remote that the imagination has no warrant for an attempt to express it even in centuries. Of all reptiles, some of the most important orders—flying reptiles, marine reptiles—have entirely passed away and left no living representative.

Now, on the principle of evolution, the most important and interesting questions are: How snakes came to be? and, What creatures may regarded as their special ancestors?

Ancient fossils throw but little light on these questions, for although the secondary period may be called the age of reptiles, snakes are not known with certainty to have had any place in it, nor yet any creatures which can be affirmed to have been the special predecessors of serpents. Nevertheless, in the latter part of the secondary deposits (which deposits ended with the chalk) a few relics have been found in Europe, and more in America (New Jersey, Alabama, and Kansas) of certain four-limbed reptiles which may have been such predecessors. They were gigantic, long-bodied animals, sometimes forty feet long, with two pairs of limbs in the form of small paddles, and were somewhat python-headed. They had jaws which could be dilated to a certain extent like those of serpents. But, as we have seen, it is not all snakes which possess this power, and since these American giants must have gained this power independently, it is difficult to see why the serpents of later times may not have done likewise. True and undoubted snakes, however, began to make their appearance in the lower tertiary rocks. They were mostly large-sized and seem all to have been non-venomous, though a fossil viper has been discovered in the middle tertiary rocks of the south of France.

Among living reptiles we can hardly expect to find the representatives of snake ancestors amid either tortoises or crocodiles, but in the very extensive order of lizards we might hope to find such, and indeed various kinds of lizards do seem to show some special resemblance to snakes, although they are probably but superficial ones.

The most obvious distinction between snakes and almost all lizards, is that the latter have two pairs of limbs, while the former have not one. But among the lizards known as scincs (whereof one was a favourite ingredient in the medicines of former days), the limbs, as we go through a long series of forms, become smaller and smaller, till they become as minute proportionally as are those of the batrachian amphiuma.* Some of these lizards have a single pair of limbs, the hinder ones, while the body is extremely elongated. Certain Australian lizards (*Delma*) have no fore limbs, while the hind limbs are only represented by a minute pair of flattened lobes,

\* See *ante*, p. 109.

K

one on either side of the body. Another allied Australian lizard has no indication whatever of limbs. The same is the case with the amphisbæna, a lizard the best known form of which is found throughout South America, where it burrows like an earthworm and has all the habits of the worm-like snakes. A nearly related kind, however, which is found in California and Mexico, though it has no hinder limbs, has a very small pair of fore-limbs, which, small as they seem, are very well formed, with minute toes. It is known as the *Chirotes*.

A lizard which in England is considered popularly to be a snake, is that known as the blind worm, which is spread throughout Europe, northern Asia, and northern Africa. Not only is it called a snake, but an adder, and sometimes a deaf adder. It is popularly reported as being deadly poison. In vain have we lectured farm labourers and even farmers on the essentially harmless and (as a slug and grub eater) beneficent nature of this small reptile. None the less have we come again and again on its mutilated form, ruthlessly cut in two, with a spade.

It is really a most gentle and inoffensive animal, which, even when roughly handled, rarely attempts to bite, while if the attempt be made, its teeth are so small, one is none the worse for it. What may help to account for the mistaken opinion that it is a kind of viper is the fact that, like that animal (but most unlike other lizards), it brings forth its young alive. In feeding, when it has seized a slug not too large for it, it passes it through its jaws till it can get one end of the slug in its mouth, when it proceeds to swallow it. Its jaws are like those of all lizards, not distendable, so that it cannot swallow either frogs or mice.

Though the belief that it is poisonous is erroneous, and though till quite lately all lizards were supposed to be non-venomous, it has of late years been found out that there may be poisonous lizards. In Central America there is a rather large lizard known as *Heloderma*, many of the teeth of which are grooved, and supplied from large spittle glands. One of these creatures was so ungrateful as to bite the hand of that distinguished American naturalist, Dr. Shufeldt. Although he at once sucked a considerable amount of blood out of the wound, yet he soon felt very severe shooting pains up his arm and down the same side of his body. The parts also rapidly swelled, and he became so faint that he fell. In two days, however, he recovered. Some of the saliva of this lizard injected into a pigeon killed it in seven minutes; and a rabbit, into the carotid artery of which some had been impelled, died in less than two minutes. It is said to affect the nervous system, including that of the heart.

But though this lizard resembles some serpents in having grooved teeth, nevertheless these teeth and its poison glands are situated in the lower jaw, whereas, as in all serpents (save one, quite recently discovered), it is only the teeth of the upper jaw which are grooved and become " poison fangs."

The structure of this venomous lizard shows us unmistakably that its poison fangs and those of serpents must have arisen independently, and poison fangs must have had a still more multiple origin, since those of the pit vipers and those of the cobras must have also had an independent origin, because their structural conditions are so diverse and the groups to which these different snakes pertain are so distinct. But such being the case, it is not impossible that the fangs of poisonous treesnakes and poisonous water-snakes also had an independent origin.

This reflection throws an important light on the possibilities of origin of the whole order of serpents, since, on the principle of evolution, that snakes must have descended from four-limbed reptiles is not to be questioned. It cannot be questioned, because the boa-snakes and the worm-snakes both bear in their bodies the rudiments of a pair of hind limbs, and therefore must be held to have descended from four-footed ancestors; but it does not necessarily follow that they all descended from the same four-footed ancestors. We have seen even in this most brief statement that different kinds of lizards (scincs, delmas, amphisbænæ, and slow-worms) have all acquired long snake-like bodies. But these kinds of lizards are not so near of kin that we need doubt but that they have acquired their snake-like forms independently, and separately parted with their limbs by persistent shrinkage. If so, why may not different sections of the order of snakes also have become elongated in body and limbless independently? Such, at least, we are persuaded, was the case with the worm-like snakes, even if the vipers, colubrines, and boas may be diverging branches from some ancient common stock. Thus we see how snakes and lizards reinforce the lesson we have had again and again impressed upon us in our successive consideration of different types of animal life, beginning with apes and opossums.

The doctrine which the student of evolution has ever to hold before his eyes to guide him in his search is the doctrine of the possibly independent origin of similar structures in very many unexpected cases.

The rattlesnake, then, is a very specially modified, exclusively American, form of pit viper. It is a poisonous snake of a special line of descent, and the most highly developed type of one primary distinct section of the whole order of serpents. It is also a member of an order of reptiles not yet certainly known to have extended back beyond tertiary times, but which is now disseminated throughout all the warmer regions of the habitable globe. It must have had four-footed ancestors, though it has not the smallest relic of a limb to boast of itself. As to what those four-footed creatures were like we cannot as yet hazard a conjecture. Since perfectly developed serpents, and even vipers, existed in tertiary times, it seems unlikely that any lizards of our own day can represent what were once the predecessors of all serpents. It is possible that in the before-mentioned python-like headed reptiles of America, we have cousins of the ancestors of snakes, if not the ancestors themselves. But this is still but a speculative hypothesis in which we cannot venture to repose anything like confidence, impressed as we are with the constantly recurring evidence before us that many similarities of organisation may co-exist without any true affinity of race and descent, accompanying such co-existence. The ancestors of the rattlesnake are, therefore, beyond our mental vision. All but enthusiastic naturalists will probably desire that their progeny may, within a moderate period, be beyond our bodily vision also,

## VI

## THE SEROTINE, OR CAROLINA BAT

THIS little brown bat has been selected as our type of all bats because it is the one only animal of the kind found in both the Old World and the New. It has, indeed, a very wide range, being found in America from Lake Winnipeg to Guatemala, while in the Old World it extends from England to Siberia, Java, and the Camaroon Mountains of Africa. It is common in all the Atlantic States, and abounds in Albany during February and March, as De Kay tells us in his "Natural History of New York." No other kind of bat whatever is found on both sides of the Atlantic.

Such small animals as the bats of temperate regions, so very rarely seen by day, and all apparently so much alike, may seem to most persons to be objects of little interest.

Nevertheless, bats are exceedingly interesting animals, as we think the reader will find to be the case. But what is a bat?

No one who has ever taken a bat in his hand and has noticed its fur, its ears, and its teeth can doubt but that it is a little beast. That the ancient Germans as well as our English-speaking ancestors saw the truth so far, is evident from the names they respectively bestowed on it —from the German name, *fledermaus*, and the old English term, *flittermouse*.

## THE CAROLINA BAT

Nevertheless bats were very often supposed to be birds. Such seems to have been the opinion of the Jews, and the "bird of darkness" is placed in Deut. xiv. 18, among the unclean ones forbidden as food: "And the stork and the heron after her kind, and the lapwing and the bat."

Aristotle, though he placed the bats among flying animals, and therefore among birds, recognised distinctly the difference in their organisation, and the same thing may be affirmed of Pliny. But in spite of this, and although Albertus Magnus, in the thirteenth century, was acquainted fully with the true nature of bats as being beasts, as also with their habit of hibernating during the cold season, we find that instead of progress a retrogression in knowledge took place after the Middle Ages.

Thus, Belon in 1557, in his "Histoire de la Nature des Oiseaux," includes bats with his birds. At the same time he was not unacquainted with the mode of their reproduction, as the following verse proves:

> "La souris chauve est un oiseau de nuict
> Qui point ne pond; ains ses petits enfante
> Lesquels du laict de ses tetins sustante
> En petit corps grande vertu reluit."

Again, almost a hundred years later on—in 1645—Aldrovandus expressed his conviction that bats were rather birds than beasts, and this in spite of his careful study of them, as proved by his beginning to distinguish different species one from another.

About a quarter of a century afterward, Ray assigned them their true place, which they have kept ever since.

But though the bat is a beast, it is a very peculiar one, and is essentially an animal of the air. All its structure is modified for flight, and it rarely descends to the ground.

In studying the turkey we saw how all a bird's structure is also modified for flight, but the modifications of bats and birds, though directed to the same end, are, as we shall see, very different modifications. Indeed, the bat's organisation, alone of existing creatures, serves to give us a good conception of the wings of those ancient flying reptiles, the pterodactyls. Its real affinities well serve to show how little mere external aspect can be trusted as a guide to fundamental relationships. The bat, as I have just said, is essentially an animal formed for aërial life above

FIG. 38.

THE CAROLINA BAT.

the surface of the ground. The mole is an animal formed for subterranean life beneath it, and the mole as rarely ascends to that surface as the bat descends to it, and its structure is so efficiently modified for most rapid burrowing that it may be said to fly through the earth as the bat flies through the air. The bat's hand, as we shall see, attains the maximum of length and slenderness, while the mole's is at a minimum of length, but is a model of concentrated power. The contrast between any animals could hardly be more complete; yet the bat and the mole share no small degree of affinity, and may be said to be zoölogical cousins.

## THE CAROLINA BAT 153

And now let us take a somewhat close look at our chosen typical form, the Serotine, or Carolina bat. It has a little rounded body about two and a half inches long, covered with a very soft fur, which Shakespeare calls "wool" when enumerating the ingredients of Macbeth's witches' cauldron. It has a small head with very small eyes, but large ears. It has a slender tail, nearly two inches long, and two pairs of limbs, extremely different both in size and structure. Its legs are of but moderate length, but disposed so singularly that the knees are bent almost backward, like our elbows.

Each leg ends in a foot with five toes, which are free (not "webbed" like those of a duck), with five claws of about the same size.

The other pair of limbs, the arms, are elongated both above and below the elbow, but the fingers are wonderfully long, and they are joined together to their tips by skin, being "webbed" like the toes of a water-fowl. But it is not only the fingers which are thus "webbed." A large expanse of skin connects them with the sides of the body, and with the legs as far as the ankles, and does not even stop there, but extends onward to the tail, which is thus connected with the two legs.

The large expanse of skin which unites the fingers and extends to the sides of the body and legs is (with its component bones, &c.) called "the wing." The part between the legs is termed the "interfemoral membrane."

If we look carefully we shall see that though the four fingers of each hand are thus bound together and support the wing membrane as the "ribs" of an umbrella support its web, each thumb is nevertheless free. Each thumb indeed stands out at a wide angle and is furnished with a very long, strong, and hooked claw.

The ear seems at the first glance to be a double organ, a very small one appearing inside the larger one. This appearance, however, is due merely to the very great development of that small prominence (called the "tragus") which in ourselves projects backward, to cover externally and so to guard the opening of the ear.

When treating of the opossum we spoke of flying opossums and flying squirrels, but no one of these creatures, any more than the flying fish or any existing reptile, really "flies."

The bat, however, flies as truly as the bird does, and in the same way—by striking the air with its fore limbs, but the mechanism is very different. When considering the turkey we noted how in birds the bones of the hand are reduced to a minimum, the fingers being both diminished in number and greatly shortened. From the brief description of the bat's wing just given we may see that in it the very opposite condition obtains.

A similar condition existed in the pterodactyls, inasmuch as they flew by means of a wing membrane sustained by the elongated bones of the hand. Nevertheless, in those reptiles it was only one finger which was thus elongated. Thus here again the similarity of their wing with that of the bat must have arisen independently.

But another independent similarity of structure, one which must have arisen at least four times over, may be noted with respect to organs which subserve the movements of the wings.

When treating of birds, we noted their almost universally "keeled" breast-bone, which, by the fact of its being keeled, affords sufficient scope for the implantation of the powerful muscles which act on the wings. Bats also require powerful muscles of the kind, and on that account have also developed a keel on their breast-bone. The same was the case with the ancient pterodactyls, and such is the case with the bat's subterranean cousin, the mole, which also requires most powerful muscles to move its short limbs as it does.

The wings being thus true organs of flight, the le s and tail together exercise a rudder-like action.

Any one who has watched the flight of bats must have been struck with the extremely rapid turns they repeatedly make—movements necessary to enable them to seize their insect food. As before said, they rarely descend to the ground, but when they do so they can crawl upon it, though in so doing they have a singularly awkward and shuffling gait. Their wings are then closed (the long fingers turned backward and lying side by side) and the animal rests on its wrists and hind feet, the body being dragged forward by the help of the strong hooked thumb nails, which also help it to climb with ease up any rough surface, even though perpendicular.

When at rest, bats usually hang suspended, head downward, by the claws of their feet, though occasionally they turn round and hang by the claws of their thumbs.

Most nocturnal beasts have large eyes, but almost all bats have very small ones. This is perhaps due to the fact that bats seem in their flight to be guided by an extraordinarily delicate sense of touch, as was long ago experimentally demonstrated by Spallanzani. He (not having any fear of anti-vivisectionists before his eyes) found that bats deprived of the power of sight, and as far as possible of smell and hearing also, were still able not only to avoid ordinary obtacles to their flight in places quite new to them, but even to pass without contact between threads which had purposely been extended in various directions across the room in which the experiments were made. This sense is believed to be due to an exceedingly delicate power of sensation possessed by the membrane of the wing—a power enabling the creature to feel by atmospheric pressure and vibration, the nearness of adjacent objects.

Certainly if the wing does possess such sensibility the great extent of its surface must intensify it to a high degree. Now, the wing is richly supplied with nerves, while the power of feeling by means of the nerves depends greatly on the amount of blood supplied to them. This we all know by the numbness we can bring easily on in any one of our fingers by tying a string tightly round its root, which causes it, as we say, to " go to sleep," a condition occasioned by depriving its nerves of their due supply of blood. The circulation of that fluid in man and beasts is brought about mainly by the rhythmical contractions of the heart, while this is aided by the elasticity of the arteries, which, though not themselves contractile, have a power, through their elasticity, of propelling the blood which is not possessed by the veins.

Now, it is a very remarkable fact that the veins in the bat's wing are positively contractile, thus serving in a most exceptional manner to propel the blood, and so, indirectly, augment such powers of sensation as the delicate membrane of the bat's wing may be supplied with.

There are probably not less than a thousand different kinds of bats, for most likely the species already collected do not amount to half those which will be eventually known to us. No less than four hundred kinds were fully described a dozen years ago by Mr. G. A. Dobson, a naturalist who has especially devoted himself to the study of these animals.

## THE CAROLINA BAT

Bats form an order of beasts primarily divided into two groups, or sections, very unequal in size. One of these comprises bats found in all parts of the world, including Europe, Northern Asia, and America.

The other group contains only the flying foxes and their allies, of which not more than about eighty species are yet known, none of which are found in America or Northern Asia and Europe.

No bats of any kind are found where neither insects nor fruit can be obtained.

Thus there are none in Iceland nor in Kerguelen's Land. They are found in most oceanic islands, even the small Savage Island, south-east of the Navigator's group, being inhabited by one kind of flying fox.

None appear, however, to inhabit the islands of the Low Archipelago or in the Galapagos group, nor have any been found in St. Helena.

The great primary division to which the Carolina bat and all American and European bats belong is made up of five subordinate groups, or families, as follows: (1) The common bats, (2) the leaf-nosed bats, (3) the Old World blood-sucking family, (4) the oblique-snouted family, and (5) the New World blood-sucking family.*

We will notice first the family of common bats, whereof more than twelve dozen different species have been already described. Though only one of these species, the Carolina bat, is common to both the Old World and the New, yet the family, as a whole, is common to both, while it ranges from 32° North latitude down to Terra del Fuego. About a dozen species of the family are found in England. The commonest of these is the pipis-

* These five families are known in science respectively by the names: (1) Vespertilionidæ, (2) Rhinolophidæ, (3) Nycteridæ, (4) Emballonuridæ, and (5) Phyllostomidæ.

trelle, which is also found throughout the whole of the northern regions of the Old World, including northern Africa. It is the first to make its appearance in England in the spring. Bats, like dormice, when winter approaches fall into a peculiar state of winter sleep, called hibernation. For this purpose they generally assemble together in large numbers, in out-of-the-way places, caverns, hollow trees, the inside of church towers, or within the roofs of outhouses, hanging head downward by the claws of their feet. During this condition the most important functions of life—breathing and the circulation of the blood—go on very slowly indeed, while the temperature of the body becomes notably diminished. From this dormant condition the pipistrelle usually rouses itself by the middle of March or soon after, and has been known even to shake off its slumbers and flit about in the middle of a bright, sunny but frosty day just before Christmas.

Its food consists specially of gnats, and as those animals often dance in the sunbeams of a winter's day in England, it is easy to understand that this little bat may then go after them. But it will eat various other insects, and even flesh, and it has been caught in a larder while making a hearty meal from a piece of meat to which it was clinging.

In confinement it has also been observed to strike down a fly with its wings and then prostrate itself over it, stretching out all its membranes to prevent the fly's escape, while it thrust down its head between its arms and secured it.

Most bats, save flying foxes, are well fitted for such food, as their grinding teeth bristle with sharp points most excellently fitted to crack the hard but brittle case which encloses an insect's body.

The flight of the pipistrelle is quick and flitting, and

it is often to be seen in the neighbourhood of ponds or streams in search of its favourite food. Its cry is exceedingly shrill, so much so, that some persons are quite unable to bear it.

Homer compares the voices of the ghosts to the cries of bats. In the 24th book of the Odyssey, 6, he says: " As when bats in a corner of a quiet cave, when one of them has fallen from off the cluster—so they (the ghosts) went along screaming."

As Pope gives it:

> "Trembling the spectres glide, and plaintive vent
> Their hollow screams along the deep descent,
> As in the cavern of some rifled den,
> When flock nocturnal bats, and birds obscene;
> Clustered they hang, till at some sudden shock
> They move, and murmurs run through all the rock.
> So cowering fled the sable host of ghosts."

Bats bring forth one or two young ones at a birth. They are born naked and blind, and are suckled much as is the human infant.

Years ago, a Mr. Daniell recorded his observations on this subject with respect to a female noctule bat, which is one of the largest species found in England. She was kept in a cage, wherein one day her owner observed that she was very restless.

The uneasiness continued for upward of an hour, the animal remaining in her usual attitude, suspended by her hind feet. On a sudden she reversed her position, and attached herself by her anterior limbs to a cross wire of the cage, stretching her hind legs to their utmost extent, curving the tail upward and expanding the interfemoral membrane so as to form a perfect nest-like cavity for the reception of the young. Into this receptacle it was born, lying on its back, perfectly destitute of hair, blind, and larger than a new-born mouse. Its hind legs and claws were remarkably strong and serviceable, enabling it not only to cling to its mother, but also to the deal sides of the cage. The dam held her baby wrapped up in the membrane of her wing, shifting it occasionally from side to side to suckle it.

Curious bats, named long-eared bats, are found both in England and the United States, though not the same species. The American species ranges from Vancouver's Island to Alabama and Florida.

FIG. 39.

THE LONG-EARED BAT.

These bats well deserve their name, for their ears are so long that they equal in length the entire trunk. They are, therefore, relatively larger than those of any other animal. They are capable of being folded up, and generally are so folded during sleep.

Speaking of this little animal, Mr. Bell tells us ("British Quadrupeds," p. 54):

"It is one of the most common British bats, and the extraordinary development of the ears, their beautiful transparency, and the elegant curves into which they are thrown at the will of the animal, render it by far the most pleasing. It is also more readily tamed than any other, and may soon be brought to exhibit a considerable degree of familiarity with those who feed and caress it. I have frequently watched them when in confinement, and have observed them to be bold and familiar even from the first. They are very cleanly, not only cleaning themselves after feeding, and at other times with great assiduity, but occasionally assisting each other in this office. They are very playful, too, and their gambols are none the less amusing from their awkwardness. They run over and against each other, pretending to bite, but never harming their companions of the same species, though I have seen them exhibit a sad spirit of persecution to an unfortunate barbastelle bat which was placed in the same cage with them. They may readily be brought to eat from the hand; and my friend, Mr. James Sowerby, had one which, when at liberty in the parlour, would fly to the hand of any of the young people who held up a fly toward it, and, pitching on the hand, take the fly without hesitation. If the insect were held between the lips, the bat would then settle on its young patron's cheek and take the fly with great readiness from the mouth; and so far was this familiarity carried, that when either of my young friends made a humming noise with the mouth in imitation of an insect, the bat would search about the lips for the promised dainty."

One of the "young friends" here referred to is now the esteemed Secretary of the Royal Botanic Society of London, and he has assured us of the truth of this anecdote.

The barbastelle bat is a kind confined to the northern regions of the Old World. It is a small bat with swollen cheeks and short ears, each containing a tragus more than half as long as the ear itself.

One found asleep in a chalk cavern in England began to wake up when brought into a warm room, when it fed readily on small bits of meat and drank water. It was fond of lying on the hearthrug before the fire, appearing to luxuriate in the warmth. It was, however, a timid animal, not at all disposed to become familiar in the way that long-eared bats will so become.

The leaf-nosed bats (2) form a family confined to the temperate and tropical parts of the Old World, from Old Ireland to New Ireland. In temperate regions they hibernate in dry and warm hiding-places during the winter, not venturing abroad in the cold. In tropical and sub-tropical countries they frequent hill regions, and many kinds are clothed with very long and dense fur. More than fifty species have been described.

These bats are very remarkable for the extraordinary folds and processes of skin which surround and decorate their noses, which appear to be excessively delicate organs of touch, no doubt capable of appreciating the proximity of objects through atmospheric pressure in an extremely high degree. This would appear to be the case both on account of the large nerves with which these organs are supplied, and also from the fact that when leaf-nosed bats are observed flying with common bats in an enclosed space they much excel the latter in dexterity.

The nose-leaf consists of three parts : (1) A more or less horseshoe-shaped fold of skin which invests the sides and front of the muzzle and includes the nostrils within its inner margin : (2) A central ridge-like process between and behind the nostrils; and (3) a membrane behind this, that either stands up vertically or extends backwards between the ears, which differ from those of the common bats, in that no sort of second ear—the tragus—stands up within them.

These bats come out later at night than the common bats, and they have especially pointed teeth to crush the dense cases of beetles on which they feed largely.

# THE CAROLINA BAT 163

When they are plentiful, some species of this family live for a great part of the year in troops counting several hundreds each and inhabiting great caverns.

After the pairing season, the females separate from the males and carry on their maternal duties in permanent "mothers' meetings." The males carry on a club life by themselves till their spouses have sent off the little ones, who can soon take care of themselves. Thereupon society life is again resumed. This cannot be said to be a universal custom, however, for one of the

FIG. 40.

THE MEGADERMA LYRA.

largest Indian species seems usually to dwell in pairs. This kind is also remarkable for being less nocturnal than most of its congeners, as it commences its flight early in the evening, and generally careers about not more than thirty feet above the ground. It seems, indeed, that it is the smaller species of insect-feeding bats which fly high, seeking small insects there to be found, while the larger kinds hawk about below, after the large beetles and other large insects which the smaller bats could not manage.

When these leaf-nosed bats are disturbed, the curious membranes on their noses are kept in constant motion, while the head is turned about in all directions as if thus to discover the cause of the disturbance.

The third family of bats I have distinguished as Old World blood-suckers, but do not by this mean to imply that the dozen species it contains all have the habit of sucking blood, but only that one typical form called Megaderma (Fig. 40) has it.

That well-known Indian observer, the late Mr. Blyth, actually captured a specimen in the act of sucking the blood, while flying, from a smaller bat which it afterward devoured. His statement is as follows (" Journal of the Asiatic Society of Bengal," vol. xi.) :

" Chancing one evening to observe a rather large bat enter an outhouse, from which there was no other egress than by the doorway, I was fortunate in being able to procure a light, and thus proceed to the capture of the animal. Upon finding itself pursued it took three or four turns round the apartment, when down dropped what at the moment I supposed to be its young, which I deposited in my handkerchief. After a somewhat tedious chase, I then secured the object of my pursuit, which proved to be a fine female Megaderma. I then looked to the other bat which I had picked up, and to my considerable surprise found it to be a small kind of pipistrelle, which is exceedingly abundant throughout India. The individual now referred to was feeble from loss of blood, which it was evident the Megaderma had been sucking from a large and still bleeding wound under and behind the ear; and the very obviously suctional form of the mouth of the Megaderma was itself sufficient to hint the strong probability of such being the case. During the very short time that elapsed before I entered the outhouse it did not appear that the depredator had once alighted; and I am satisfied that it sucked the vital fluid from its victim as it flew, having probably seized it on the wing, and that it was seeking a quiet nook, where it might devour the body at leisure. I kept both animals separate till next morning, when, procuring a convenient cage, I first put in the Megaderma, and after observing it for some time, I placed the pipistrelle with it. No sooner was the latter perceived than the other fastened upon it with the ferocity of a tiger, again seizing it behind the ear, and made several efforts to fly off with it; but finding it must needs stay within the precincts of the cage, it soon hung by the hind legs to one side of its prison, and after sucking its victim till no more blood was left, commenced devouring it, and soon left nothing but the head and some portions of the limbs."

The members of this small family are confined exclusively to the warmer parts of Africa and Asia, from Egypt to Celebes. They have a very conspicuous noseleaf and large ears, medially united to each other above the head, and each with a large tragus within.

The oblique-snouted family of bats is very large, sixty-three species having been already described a dozen years ago. It has representatives in both hemispheres.

Seven genera (with thirteen species) are peculiar to America, five are peculiar to the New World, while two are common to both.

These bats have no nose-leaves, and the faces of some of them remind us of pug dogs. The tail projects freely beyond the short interfemoral membrane. Many of them have narrow wings, and some are very naked.

The most curious form (*Cheiromeles*) from the Malay region, has a very thick skin, almost naked, while its great toe is very large and separated from the others. A curious fold of skin on the breast and sides of the body serves as a cradle for the baby. Such nursing pouches are probably absolutely necessary for the preservation of the young, which otherwise could scarcely maintain their hold on the naked body of the mother during flight.

It is interesting to find these pouches developed in

both the male and the female, for their presence in the former suggests the idea that, where the young are born together, the male may relieve the female of one of them.

The fifth and last family of the larger primary section of the order of bats is that which I have distinguished as the New World blood-suckers. It indeed is confined exclusively to South and Central America, save one species, which is said to extend up to Bermuda and South Carolina.

There are from sixty to seventy species, among which the renowned vampires are included. All of them possess nose leaves, but, unlike the Old World nose-leaf bats, they also have a well-developed tragus within the ears, and also rather large eyes. It appears that only one or two of the family are really blood-suckers, and those kinds which in science are specially distinguished as vampires appear to be insect-eating bats.

All sorts of exaggerated accounts were given, although some old observations which some discredited are now found to have been justified. D'Azara affirmed that they would sometimes bite the wattles and crests of fowls while asleep, and suck their blood. The fowls, he said, generally die of this, as gangrene is engendered by the wounds. He adds:

"They bite also horses, mules, asses, and horned cattle, usually on the shoulders, buttocks, or neck, as they are better enabled to arrive at those parts from the facilities afforded them by the mane and tail. Nor is man himself secure from their attacks. On this point I am able to give a very faithful testimony, since I have had the ends of my toes bitten by them four times while I was sleeping in the cottages in the open country. The wounds which they inflicted, without my feeling them at the time, were circular, or rather elliptical."

The late Mr. Darwin was fortunate enough to be able not only conclusively to prove the truth of this bloodsucking habit, but also to capture an individual in the act, and to make sure exactly what species it was. It is that known as "Desmodus," a form which ranges from Mexico to Chili.

FIG. 41.

THE VAMPIRE *(Desmodus)*.

Speaking of horses, Mr. Darwin tells us ("Voyage of H.M.S. Beagle," vol. i. p. 22) that this animal

"is often the cause of much trouble by biting horses on their withers. The injury is generally not so much owing to the loss of blood as the inflammation which the pressure of the saddle afterwards produces. The whole circumstance has lately been doubted in England; I was therefore fortunate in being present when one was actually caught on a horse's back. We were bivouacking late one evening near Coquimbo, in Chili, when my servant, noticing that one of the horses was very restive, went to see what was the matter, and fancying he could distinguish something, suddenly put his hand on the beast's withers and secured the bat. In the morning the spot where the bite had been inflicted was easily distinguished from being slightly swollen and bloody. The third day afterward we rode the horse, without any ill effects."

The structure of this bat is wonderfully modified in harmony with its habits. The special modifications are of two kinds—first, the form of the teeth; and secondly, that of the stomach.

We have already called attention to the fact that the back teeth of most bats bristle with sharp points. They are also proportionately of good size, while the front or cutting teeth are very small indeed. In this bat, however, the back teeth are reduced to a minimum both in size and number, being quite rudimentary; at the same time, the two middle cutting teeth of the upper jaw are of good size and provided with sharp cutting edges, like lancets.

They are thus admirably fitted to make the small puncture which the animal requires to make in order that it may obtain its needful nourishment.

The stomach presents us with a structure quite unique in the animal kingdom. The stomachs of most bats yet noticed are more or less rounded structures, not extending far either right or left from the spot where the gullet enters it. It is the part on the left of the gullet's entrance which is the more digestive portion, and in some animals it is very much enlarged and subdivided.

The part on the right of the gullet is large in such creatures as sheep and oxen, and it receives the fresh-cropped herbage before digestion begins. In this curious bat the left or more digestive part of the stomach is reduced to a mere rudiment—the highly nutritious food (blood) requiring very little digestion. A capacious cavity is, however, needed for its reception, and accordingly the part of the stomach on the right of the gullet is not dilated into a mere capacious sack, as in the sheep, but is drawn out into an enormously long and wide tube, capable of containing a large quantity of fluid. So greedy, however, is this bat that it will continue to suck blood after its capacious intestines are entirely filled with it, the blood first drawn escaping from the latter while fresh blood is being sucked in by the mouth.

It is now time to notice the other great primary section of the order of bats—namely, the flying foxes.

Of these, as before said, about eighty species are known, none of them being American. They range from Asia Minor and Egypt through Africa and Asia to Australia, the Fiji, and Duke of York's, and Navigator's Islands and New Ireland. None are found in Tasmania or New Zealand. Among these are found the largest of all bats. The body may be a foot long and the outstretched wings measure five feet across. They are also the most brightly coloured and the most varied in tint. Only in one species is there a long tail; in all the others it is short, or may be entirely absent. The first finger of the wing generally bears a claw. These bats feed on fruit and not on insects, and therefore their teeth, instead of bristling with sharp points, are smooth, save that they each bear a longitudinal furrow.

The stomach is not rounded, as in most bats, but elongated. Its elongation, however, is just the opposite of that of the blood-sucking desmodus, and it is the left or digestive portion of the organ which is elongated.

The largest of these bats is that known as the kalong. It inhabits the Indian Archipelago, extending from the Andaman and Nicobar Islands to the Philippines and Timor.

FIG. 42.

THE KALONG.

In the lower parts of Java it is very common, and lives in troops, which do not appear to visit the more elevated districts.

## THE CAROLINA BAT. 171

Numerous individuals select a tange tree for their resort, and, suspending themselves by the claws of their hind limbs to the naked branches, often in companies of several hundreds, afford to a stranger a very singular spectacle. A species of fig-tree, often found near the villages of the natives, affords them a favourable retreat, and the branches are sometimes covered with them. They pass the greater part of the day in sleep, hanging motionless, ranged in succession, with the head downward, the wing membrane contracted about the body, and often in close contact; looking like fruit of uncommon size suspended from its branches. In general these societies preserve a perfect silence during the day, but if they are disturbed, or if a contention arises among them, they emit sharp piercing shrieks, and their awkward attempts to extricate themselves when oppressed by the light of the sun exhibit a ludicrous spectacle. Soon after sunset they gradually quit their hold, and pursue their nocturnal flight in quest of food. They direct their course by an unerring instinct to the forests, villages, and plantations, occasioning incalculable mischief, attacking and devouring indiscriminately every kind of fruit, from the abundant and useful wood-nut, which surrounds the dwellings of the meanest peasantry, to the rare and most delicate productions which are cultivated with care by princes and chiefs. By the latter, as well as by the European colonists, various methods are employed to protect the orchards and gardens. Without such precaution but little valuable fruit would escape the ravages of the kalong. They may be observed as soon as the light of the sun is gone. Then the bats may be seen to follow each other at small but irregular distances, and this succession continues till darkness obstructs the view. The flight of the kalong is slow and steady, pursued in straight line, and capable of long continuance. The hunting of these bats forms occasionally an amusement during the moonlight nights. Each is watched till it descends on a fruit tree, and then a discharge of small shot will bring it to the ground. Four or five specimens may thus be obtained in an hour.

Most of the flying foxes inhabit trees, but some also are found in caverns with various other species of bats.

Mr. Pryor (a corresponding member of the Zoölogical Society of London) had a curious experience respecting bats in caves when he explored the caverns of North Borneo, which are inhabited by the swift, which make the edible nest so much prized by the Chinese. He tells in "Proceedings of the Zoölogical Society," 1884, p. 534:

"After a rest I ascended the cliff about 400 feet. The ascent is quite perpendicular. In many places ladders are erected, and in others the water-worn surface of the limestone gives a foothold. On the ascent I noticed many orchids, begonias, ferns, and mosses, I had not seen elsewhere. My collector caught a snake I believe to be an *Elaphis*, certainly the most beautiful Colubrine I have seen, white and light grey. The Malays said it was very destructive to the swifts, and also that it was poisonous; to convince them it was not, I allowed it to bite me. At this point I found myself at the mouth of a cave mamed Simud Putih—*i.e.*, the White Cave. The entrance is about 40 feet high by 60 feet wide, and descends very steeply, widening out to a great size, and having a perpendicular unexplored abyss at its furthest point. This cave is used by the nest-gatherers as their dwelling-place, and at the entrance are their platforms of sticks, one of which was placed at my disposal by the head man; it is also the cave by which the great body of the swifts enter. Immediately outside it is a great circular opening leading sheer down into Simud Itam; this is one of the two openings mentioned as giving light to that cave, and is the entrance most in use by the bats. As soon as I had unpacked and settled down on my platform I sallied out to find the material from which the birds make their nests, as my previous experience is that birds do not as a rule travel far for the bulk of the material they use. I was speedily successful in my search. It is a fungoid growth which incrusts the rock in damp places, and when fresh resembles half-melted gum tragacanth; outside it is brown but inside white, and little if any change in its consistency is effected by the bird; the inside of the nest is, however, formed by threads of the same substance which are drawn out of the mouth in a similar way to that of a caterpillar weaving its cocoon.

"The Malays told me to be sure and return to Simud Putih at five o'clock, as I should then see the most wonderful sight in all Borneo—the departure of the bats and the return to roost of the swifts. I accordingly took a seat on a block of limestone at the mouth of the cave; the surface of the coral of which it is composed is quite fresh-looking, notwithstanding that it must have been many ages in its present position, several hundred feet above sea level. Soon I heard a rushing sound, and, peering over the edge of the circular opening leading into Simud Itam, I saw columns of bats wheeling round the sides in regular order. Shortly after five o'clock, although the sun had not yet set, the columns began to rise above the edge, still in a circular flight; they then rose, wheeling round a high tree growing on the opposite side, and every few minutes a large flight would break off, and, after rising high in the air, disappear in the distance; each flight contained many thousands. I counted nineteen flocks go off in this way, and they continued to go off in a continual stream until it was too dark for me to see them any longer. Among them were three albinos, called by the Malays, the Rajah, his son and wife.

At a quarter to six the swifts began to come into Simud Putih. A few had been flying in and out all day long, but now they began to pour in, at first in tens and then in hundreds, until the sound of their wings was like a strong gale of wind whistling through the rigging of a ship. They continued flying in until after midnight, as

I could still see them flashing by over my head when I went to sleep. As long as it remained light I found it impossible to catch any with my butterfly net, but after dark it was only necessary to wave the net in the air to secure as many as I wanted. Nevertheless, they must undoubtedly possess wonderful powers of sight to fly about in the dark in the deepest recesses of their caves and to return to their nests, often built in places where no light ever penetrates.

Shortly before sundown a pair of kites made their appearance, and, taking their station over the bat chasm, would every now and then sweep down into the thick of the bats, generally securing a victim every time. I shot both these marauders, which proved to be *Haliastur indus*, a very beautiful but common bird. There were also several specimens of a hawk, working away on the bats in a very business-like manner, and woe betide the unfortunate bat singled out from its flock and put in chase! The way these hawks took the bats one after the other was astonishing, and strongly reminded me of a man eating oysters. I shot several of these hawks, but only secured one, the others being lost over the side of the cliff. It proved to be the rare *Machirhamphus alcinus*, remarkable for the size of its gape and its small beak, both of which very much resemble those of the swifts. Its habits in taking its prey are also similar, the swift catching and swallowing its food while on the wing in the same way as this hawk does.

Arising before daylight, I witnessed a reversal of the proceedings of the previous night, the swifts now going out of Simud Putih and the bats going into Simud Itam. The latter literally "rained" into their chasm for two hours after daylight. On looking up, the air seemed filled with small specks, which flashed down perpendicularly with great rapidity and disappeared in the darkness below. . . . . I secured many specimens of the bat, and found them to be all of one species. The wings are very long and narrow, and it is a very swift flyer. I noticed a few specimens of a swallow and also some very large bats at the mouth of the cave. These large bats were, of course, some kind of flying foxes.

## THE CAROLINA BAT

We have now noticed the main groups of bats which inhabit the world in our day; but we know little indeed of bats which inhabited it in earlier epochs. The

Fig. 43.

THE COLUGO.

oldest known remains are but fossils found in tertiary deposits, and they offer us no startling revelation.

Some form of existing beasts which are now distinct enough (such as the ox, the pig, and the horse) were preceded in early tertiary times by others which were more or less intermediate in structure. This is not the case as regards bats. Bats, as soon as they appear at all, appear as thoroughly and as perfectly organised as are those bats living among us now. And living bats are separated from all other beasts in a very marked manner. They constitute an order by themselves, and this fact, together with the various others we have been able to set down, may enable the reader to answer the question, "What is a bat?" in a reasonable manner.

But the questions, How bats came to be? What was the origin of the bat? we are by no means able to answer. We cannot say what creatures may have been the bat's genetic predecessors, or at what epoch the bat first appeared, save that it was before the deposition of the tertiary rocks.

There is one animal, found in Singapore and Borneo, which has been supposed to show some affinity to bats. This is the colugo, or, as it is sometimes called, the flying lemur (Fig. 43).

It has its fingers webbed, while a membrane extends on either side between the arms and the legs, and from the legs to the tail. So far it is like a bat, but its fingers are not elongated and its toes are webbed, while those of the bat are not. Moreover, though it takes long jumps through the air and may be able somewhat to guide its flight, it certainly does not truly fly. We cannot therefore regard this animal as exhibiting any indication of the source of the bat tribe.

We must, it seems, wait for more light from the stores of yet undiscovered fossils which the earth contains.

# VII

## THE AMERICAN BISON

THE American bison, universally known in the United States as the buffalo, is one of the grandest of all the creatures which "divide the hoof and chew the cud." It has a large head, with shortish, rounded horns, and there

FIG. 44.

THE AMERICAN BISON.

is a sort of hump over the shoulders, due to the height of the withers, which (as in a specimen in the Museum at Washington) may be five feet eight inches high. The hind-quarters, however, are low and comparatively weak. The body, generally, is covered with short, more or less

M

dark-brown woolly fur. Long, darker hair on the head hides the eyes, ears, and bases of the horns, while a shaggy coat or mane clothes the neck, withers, shoulders, and thence downward to the knees. There is a long beard beneath the chin, and the tail is tufted at its end. The length of the head and body to the root of the tail may exceed ten feet by two or three inches.

The buffalo should be a very interesting animal to all American citizens on account of the great danger which exists of its becoming utterly extinct. Only thirty-one years ago they still numbered several millions, more than five millions at the least, whereas in 1889 there were but some twenty individuals in Texas, a few in Colorado, Wyoming, Montana, and Dakota, and two hundred preserved by the Government in the Yellowstone National Park. We have, however, recently been assured that some private individual citizens in the United States are trying to preserve and propagate the buffalo. Canada, which now exhibits such interesting examples of political and social "survival," has been practically conservative as regards the bison, since it appears that some 500 individuals of a race known as the wood bison still survive there. We trust that all lovers of Nature will have cause to be grateful to the Fiftieth Congress, which at its last session voted $200,000 for the establishment of a National Zoölogical Park on a grand scale in the District of Columbia, with the intention that American quadrupeds now threatened with extermination should enjoy a luxurious captivity, when it is hoped they may breed. If this project be duly carried out, we may be confident that the bison will breed there, since it has been known to breed in captivity as long ago as 1786.

Strange to say, this animal seems to have been first seen by Europeans, not in a wild state, but preserved in a menagerie. It was thus seen by Cortez in 1521, when he reached Anahuac, where the Mexican king, Montezuma, maintained a collection of wild animals, among them a bison, which must have been brought a distance of 400 miles at the least. It was first met with wild, in 1530, by Alvar Nuñez Cabeza, in south-eastern Texas, while an English traveller, Samuel Argoll, saw it somewhere near Washington in 1612. At one time it existed in enormous quantities, the prairies being absolutely black with them as far as the eye could reach. Col. Dodge tells us in his "Plains of the Great West" that, even so late as May 1871, he drove thirty-four miles in a light waggon from old Fort Zara to Fort Larned on the Arkansas. For at least twenty-five miles of this distance he passed through one immense herd, composed of countless smaller herds of buffalo then on their journey north. The whole country appeared one great mass of buffalo, moving slowly to the northward; it was only when actually among them that it could be ascertained that the apparently solid mass was an agglomeration of innumerable small herds of from 50 to 200 animals. Its range once extended over about a third of the whole of North America. In some places, as in Georgia, it almost reached the Atlantic coast, extending thence westward through the Alleghany Mountains and forests to the prairies of the Mississippi—always its special home—and southward to north-eastern Mexico; also across the Rocky Mountains to New Mexico and Utah, and northward to the Great Slave Lake. Readers who may be interested to know further details about the bison and its approach to extermination are referred to J. A. Allen's admirable monograph, "The American Bison, Living and Extinct;" and to William T. Hornaday's work, "The Extermination of the American Bison," published at

Washington under the auspices of the Smithsonian Institution.

In consequence of the settlement of the country by Europeans the area inhabited by the bison was gradually contracted until, about 1840, one mighty herd occupied the centre of its former range. The completion of the Union Pacific Railway in 1869 divided this great herd into a southern and northern division, the former comprising a number of individuals estimated at nearly four millions, while the other contained about a million and a half. Before 1880 the southern herd had practically ceased to exist, while the same fate threatened the northern one in 1883, till in 1889 the species became reduced to the numbers before given for that year.

But to know all about the form and structure, the habits and distribution of this animal, will go a very little way towards enabling us to answer a very important question, with asking which we might very well have begun this article—the question, "What is the animal which Americans know as the buffalo?" To answer this we must learn something of its relations to all other animals, beginning, of course, with those which are most like it and may be supposed to be very closely akin to it. An extinct bison from the pleistocene rocks of Texas, has been distinguished as the "broad-fronted bison." There is, however, one species still living—the auroch. It is not only confined to the Old World, but is now nowhere to be met with except in the primeval forests of Lithuania, Moldavia, Wallachia, and the Caucasus, where it is artificially preserved. Formerly it doubtless ranged over a large portion of Europe. It is very like the American kind, but is slightly larger, with more powerful hind-quarters; the fore part of the body, however, is not so massive, nor is the mane so luxuriant.

The European aurochs and the American bison thus form a pair of species which are separated off from all other animals by certain details of their structure. Their nearest ally appears to be the Asiatic animal known as the yak, a beast which in a wild state inhabits Chinese Thibet. The yak differs from the bison in not having a mane, but it has something the appearance of a mediæval knight's caparisoned horse with flowing drapery on either side. This is represented in the yak by a fringe of long hair

FIG. 45.

THE YAK.

hanging down from the shoulders, flanks, and thighs nearly to the ground, while the tail bears a wonderful mass of long, silky hair. Tame yaks are used as beasts of burden, and they are very serviceable for traversing the high, desolate regions of Thibet, and would be much more so but for their requiring grass as food and refusing corn. They are often crossed with domestic cattle, and the white tails of such half-bred animals are much valued in India, where they are known as "chowries," and used as fans to drive away flies and other insects.

The true buffaloes come very near the bison in form and structure, but they have more or less flattened horns, which incline upward and backward, the tips curving inward. The Indian buffalo has been domesticated in Egypt and southern Europe. The wild animal is a huge beast with enormous horns, and frequents swampy, moist districts. There are two African species, and a small kind, known as the anoa, is found in Celebes. Three

Fig. 46.

THE CAPE BUFFALO.

other kinds of large, ruminating animals lead from the bisons and buffaloes to the true oxen. These are the gaur, the gayal, and the banteng. The gaur is the largest, being fully six feet high at the withers. It is found in all the large Indian forests south of the Himalayas, and is known to sportsmen as the Indian bison. It is very shy, and has never been domesticated. The gayal has not yet been found wild, though semi-domesticated individuals occur in Spain and parts adjacent. Its light-coloured legs, which look as if the animal had on white stockings, give the creature a very singular appear-

## THE AMERICAN BISON 183

ance. The banteng is smaller than either the gaur or the gayal, and is found in Burmah, Java, Bali, and Lombok. The true oxen are represented by two species. One of these consists of those handsome beasts, the humped cattle of India. The other includes the domestic oxen of Europe and America, together with the herds of wild cattle which are still preserved at Chillingham and some other British parks. Wild oxen were abundant in

FIG. 47.

THE MUSK OX.

European forests in the days of Julius Cæsar, who gave them the name of the urus, and described them as equalling the auroch in size.

All the species yet noticed may be spoken of as bovine animals or "oxen," in the widest sense of the term. Between them and the goats and sheep, or caprine creatures, stands an intermediate form known as the musk ox. This is of about the size of a small Welsh or Scotch ox, and covered with thick, brown, matted, and curly hair. It goes in herds of from twenty to eighty and a hundred individuals, among which only two or three full-grown males are to be found. When frightened they collect together as a flock of sheep will do, and similarly follow the leader of the herd. In Pleistocene times this animal ranged over northern Siberia, Germany, France, and England, but in the present day it is confined to the more northern parts of North America and the shores of the Arctic Sea. It extends through the Parry Islands and Grinnell Land to North Greenland, and is found in Sabine Island. Sir J. Richardson tells us that when the animal "is fat, its flesh is well tasted and resembles that of the caribou, but has a coarser grain. The flesh of the bulls is highly flavoured, and both bulls and cows, when lean, smell strongly of musk, their flesh at the same time being very dark and tough." But the observations of Major Feilden show that the flavour of its flesh varies much from some unknown cause, independently of age, sex, or the season of the year.

The group of goats and sheep is a numerous one. Among the former (which includes some dozen species), may be mentioned the ibex, the markhoor, and the thar. All the goats are exclusively confined to southern Europe and northern and central Asia. There are about twelve kinds of sheep, whereof the big-horn, or mountain sheep, is the only kind which is naturally an inhabitant of the New World, where it was never domesticated. Wild sheep are all but exclusively confined to central Asia, but the aoudad inhabits the mountains of northern Africa, and the moufflon, Corsica and Sardinia. The bharal is found in the Himalayas and the argali in Mongolia. Sheep are naturally dwellers in mountainous regions, and none voluntarily take to forests, swamps, or level plains. A fossil sheep resembling the argali has been found in England, otherwise the sheep is not

## THE AMERICAN BISON

supposed to have been naturally an inhabitant of the British Isles.

A number of other forms must here, for convenience' sake, be grouped together as antelopes. Like all the kinds yet noticed they have hollow horns supported on bony cores, and neither the horns nor the cores are ever shed. They are mostly Old World forms, while none are met with in South or Central America. Among the

FIG. 48.

THE ROCKY MOUNTAIN GOAT.

few American species is that known as the "Rocky Mountain goat," which inhabits the northern part of California, while the chamois is an inhabitant of Europe from the Pyrenees to the Caucasus, ascending to the limit of perpetual snow. A multitude of species come from southern Africa, such as the hartebeest, the blessbok, the bontebok, the springbok, the curious gnu with its ox-like head and delicate feet, the duilkerbok, Salt's antelope, the royal antelope, the rehbok, the waterbok, the singsing, the reitbok, the palla, the sable antelope, the blaubok, Baker's antelope, the gemsbok, the leucoryx, the beautiful harnessed antelope, the kudu, and the eland. Among Asiatic forms may be mentioned the curious and very exceptional four-horned antelope of India, the black buck of Bengal and Malabar, the saiga of the steppes of central Asia, the chiru of Turkestan, more than twenty species

FIG. 49.

THE HARNESSED ANTELOPE.

of gazelles, and the elegant, short-horned nilghai. Ranging from eastern Africa to western Asia the addax and the oryx are to be found.

All the animals yet noticed have horns which are unbranched, and when once developed are not shed. North America, however, possesses a very strange form known as the prong-horned antelope, which is exceptional in two respects: (1) Each of its horns gives off a branch extending forward, although the bony core is unbranched.

(2) At intervals each horn is cast, after which a new one is formed on the bony cores beneath it. This species terminates our list of creatures with horns formed on essentially the same type as those of the animal known in the United States as the buffalo. Thus the American bison is a very exceptional hollow-horned ruminant;

FIG. 50.

THE PRONG-HORNED ANTELOPE.

in other words, it is a member of a great group—a family —which contains, together with the oxen, all the sheep, goats, and antelopes.

Another great group of allied animals consists of the deer, whereof the magnificent North American deer, the wapiti, may be taken as a type. None of the deer have outgrowths of true horny matter deposited on a bony support. Instead of that they have antlers—that is, outgrowths of bone from the head which, while they are growing, are covered with soft, sensitive skin, richly supplied with blood. When once completely formed the supply of blood ceases, and the skin, becoming dry, wears off, leaving the bone of the antler naked and bare. After

FIG. 51.

THE WAPITI.

a time the antler becomes detached near its base and falls off, the part left, or pedicle, serving to develop the antler which next succeeds. Antlers may be simple and straight, but they mostly send off branches or snags, as is well seen in the magnificent antlers of the wapiti. There are no antlers in two species, while in the reindeer they are present on both sexes. In all other deer only the males

## THE AMERICAN BISON 189

possess them. The two forms devoid of antlers are the Chinese water deer and the musk deer, the latter being an inhabitant of central and eastern Asia, where it dwells in high elevations. Among the other deer may be mentioned the muntjacs of south-eastern Asia, with very short antlers borne on long pedicles, the Sambur deer, hog deer, swamp deer, the axis, the fallow deer, the Cashmir deer, and the sika, all from Asia; the red

FIG. 52.

THE MUSK DEER.

deer of Europe, western Asia, and northern Africa, the roebuck of Europe and western Asia, the moose and caribou of the extreme north of the Old and New Worlds, the wapiti, the Virginian, Mexican, and mule (or large-eared) deer of North America, and a few South American forms known as brockets. With these two great groups (1) the hollow-horned and (2) the antler-bearing animals, are to be classed three small and exceptional groups, only one of which has horns of any kind. The first of these three groups consists only of the giraffe, a creature now confined to the Ethiopian region of the earth's surface, though in Pliocene times it existed in Greece, Persia, India, and China. It has a pair of short, bony processes on the head, coated with hairy skin, and something like

FIG. 53.

THE CHEVROTAIN.

the pedicles of deer, and there is also a median bony protrusion between the eyes, like a low, blunt, third kind of horn. Another of the three exceptional groups above mentioned is that composed of the chevrotains, which are the smallest of all ruminating animals. Three species of these come from the Malay Peninsula, another from Ceylon, while a fifth and larger kind is found on the banks of streams in western Africa. These chevrotains are very often called musk deer, a most unfortunate mistake, as they have no special affinity whatever with the true musk deer.

The last group of ruminating animals consists of the camels and the llamas, whereof the former are Asiatic, while the latter are now confined to the western side of South America. The zoölogical position of the bison thus appears to be situated near one end of a vast series of ruminating animals, at the opposite end of which the camel has its place. These may be represented in tabular form as follows:

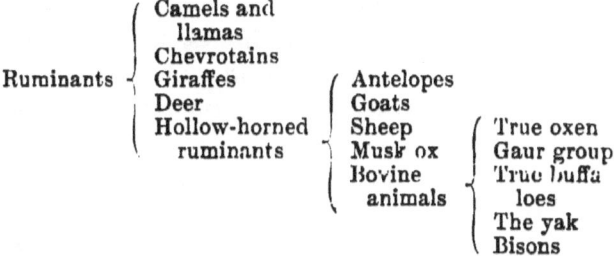

But the great group of ruminants, considered as a whole, forms one section of a yet larger group of animals which are characterised by having an even number of toes on each hind foot. As the end of each toe supports, or is enclosed in, a very large nail or hoof, the whole group may be distinguished as "even-toed hoofed beasts."

The other section of this whole group consists of the hippopotamus and the swine of the Old World and the peccaries of the New World, and all these together constitute a group which may be distinguished as the "non-ruminating even-toed hoofed beasts." But this whole group of "even-toed hoofed beasts" thus divided into (1) "ruminating" and (2) "non-ruminating" sections, has opposed to it another group of "odd-toed hoofed beasts," that is, beasts the number of whose hind toes, made use of in locomotion, is either one, as in the horse, or three, as in the rhinoceros and the tapirs.

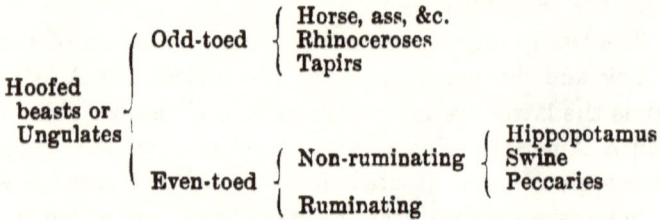

Thus we see that hoofed beasts, or ungulates, may be odd- or even-toed, that the even-toed may be ruminating or non-ruminating, and that the ruminating may be allied to camels, chevrotains, giraffes, deer, or hollow-horned cattle, and the bison is an exceptional form of the hollow-horned portion of the ruminating even-toed hoofed beasts.

What relation, then, do hoofed beasts as a whole bear to all other beasts?

By understanding the answer to this question we shall come sufficiently to comprehend what a bison really is. We will begin by adverting to existing animals only, since if, as the most popular theory of evolution teaches, all existing kinds have been evolved, by minute modifications, from pre-existing kinds, then, if we could know the whole, we should be unable to draw any fundamental distinctions at all between any groups of animals whatever, which would form an immense mass melting into each other by insensible gradations. Now, of the existing orders of mammals there are very few which show any affinity to the hoofed beasts, or ungulates. Certainly we can find no evidence of it among existing apes, lemurs, bats, insect-eaters, flesh-eaters, rodents or whales. The kangaroos among marsupials have been supposed to bear a certain affinity to the ungulates or hoofed beasts, but this is no real affinity, but only a certain degree of analogical resemblance. In other words, it can have nothing to do with essential affinity or any special relationship of descent. Again, a certain resemblance between such creatures as sloths, ant-eaters, and armadillos on the one hand, and hoofed beasts on the other, may be imagined to exist, on account of the very large claws which invest the ends of their digits and which might be considered as comparable with hoofs. But such a resemblance would be but fanciful; for sloths and their allies have no real affinity to either the odd- or even-toed ungulates. There are, however, two animals which have been supposed to possess an exceptional relationship to the ungulates—namely, the elephant and the hyrax, or cony of Scripture.

But before considering the right which these two very different kinds may possess to claim a real affinity to the ungulates—that is, to all kinds of cattle—we must consider a few structural characters bearing upon the distinctions which exist between hoofed beasts and other beasts and between different groups of ungulates themselves. Now, in the first place, the fully developed foot of a beast—like the foot of man—consists of five toes, the end of each being furnished with a nail or claw. Each toe is supported by three bones attached, end to end, within it, except the great toe, which has but two. These five series of toe-bones are each respectively attached to the end of one of five longer and stronger bones, which lie along the middle of the foot, and so may be called middle-foot bones. They are attached by their other ends to a group of short bones, which may be called ankle bones, the hindmost of which forms the heel. The fore-feet of beasts and man's hands are both formed on the same fundamental plan as is the hind foot. Each of the four fingers is supported by three bones, attached end to end, while the thumb, like the great toe, has but two. These five series of finger-bones are each respectively attached to the end of one of five longer and stronger bones which lie along the middle of the hand, and so may be called middle-hand bones. They are attached by their other ends to a group of short bones, which are the wrist bones. Now, very many beasts, like the bear, agree with man in walking with the entire sole, or plant, of the foot directly applied to the ground, and such are, therefore, called plantigrade. Others, like dogs and cats, walk on their toes only, and keep the soles of the feet and the parts which answer to the palms of our hands raised up from the ground. Such animals are called digitigrade, because they walk on their digits, which is a common term used to denote equally both fingers and toes. Now, no ungulates place the plant of the foot on the ground, while almost all of them do not rest even on their toes, but only on the very tips of their toes, for they walk on their hoofs—that is to say, on the great nails which encase the terminal bone of each finger and toe, or, in other words, the terminal bone of each digit. It is only in the camels and llamas, which are digitigrade and which have coarse nails instead of true hoofs. No ungulate has either a thumb or a great toe, while the number of digits may be reduced to one for each limb.

In the odd-toed ungulates the bones of the foot are developed symmetrically on either side of an imaginary axis passing down the middle of that digit which corresponds to our middle finger and middle toe. Such is the case even with the tapir, which has four toes to each hind foot, though the outermost is not reckoned a functional one, the animal resting on but three. The rhinoceroses

## THE AMERICAN BISON

have only three toes to each foot, while the horse, as also the ass, the zebra, and the quagga, has but one. It would be quite a mistake to suppose that the horse's foot consists of a hoof which is not divided; it consists of but a single digit, and the horse's four feet answer to the tips of our two middle fingers and the tips of our two middle toes, and nothing else. The enormous toe which supports the horse contains three bones similar, except as to size and details of form, to those of our own middle finger and toe; and similarly again, each fore digit is supported by a single middle-hand bone, and each hind digit is supported by a single middle-foot bone, at the top of which are the small bones of the wrist and ankle respectively. Thus, what is commonly called the "knee" of a horse is truly its wrist, and what is called the "hock" of a horse is really its heel. Though, however, the horse has thus but a single digit and middle-foot bone to each limb, there is, nevertheless, a slender bone on each side of it, though it bears no digit. These two slender bones, called by veterinary surgeons splint bones, are the only rudiments of the second and fourth toes.

Such are the conditions found among existing odd-toed ungulates. With the even-toed group it is otherwise. There are always two toes equally developed and corresponding with our third and fourth toes, and the line of symmetry of the foot passes down between them, instead of along the middle of the third digit, as in the odd-toed hoofed beasts. Two other digits, corresponding with our second and fifth digits, are very often present, but they are always smaller than the third and fourth ones. They are at least so in the hippopotamus. In the swine they do not reach to the ground.

The only American non-ruminating, even-toed ungulates, the peccaries, present two exceptional characters in the structure of their feet. In the first place, the hind-foot has three digits, yet this does not make it an odd-toed ungulate, because only two are functional and the line of symmetry passes down between them, as in the other even-toed hoofed beasts. The fore-foot is as in the swine. The other exceptional character is the median union of the upper portions of the two middle-foot bones into a single solid structure or cannon bone. This union exists in all the ruminants except the African

FIG. 54

THE COLLARED PECCARY.

water chevrotain. The camels and llamas have but two toes to each foot, but the other ruminants have generally the two lateral digits more or less represented. In the bison there are only minute representations of the two lateral digits, with the addition, in each fore foot, of a short rudiment of each corresponding middle-hand bone, but quite separate from the digits to which they correspond.

Beasts in general have, like ourselves, three kinds of teeth—cutting teeth, in front; eye teeth, called canines because they are so big in the dog; and grinding teeth. Of the first there are commonly six above and six below; of the second, one on each side above and one below; of the third kind, seven above and seven below on each side. This is the number in the swine, where the eye teeth become tusks, and it is almost the same in the hippopotamus and peccaries. The camels and llamas have fewer grinders, and only two cutting teeth in the upper jaw. In all the rest of the ruminants there are no cutting teeth in the upper jaw, which is clothed with a callous pad against which the lower cutting teeth and eye teeth (which are shaped like the cutting teeth) bite. In most ruminants the upper eye teeth are wanting, as well as the upper cutting teeth, the whole front of the upper jaw being absolutely toothless. But in the camels, llamas, and chevrotains they are present, and in the musk deer, muntjacs, and Chinese water deer they are very large and prominent in the males. Among the odd-toed ungulates the tapirs are toothed, as to number, like the swine. The horse has one grinding tooth less on each side below, while the cutting teeth have an absolutely peculiar structure, each having a deep indention or pit on the middle of its cutting surface. This pit, getting filled with particles of food, becomes of a dark colour, and constitutes what is known as the "mark." The presence of this "mark" shows that a horse has not exceeded a certain age, since when the tooth has worn down beyond the extent of the inflection, the "mark" becomes thereby obliterated. The rhinoceroses have the same number of grinding teeth as swine have, and sometimes have no teeth whatever in front of them, though they may have four cutting teeth both above and below.

For digestion, every beast possesses a stomach and intestine, and the latter tube has generally projecting from it on one side, at a certain point, an enlargement called a "cæcum." Now, all the odd-toed ungulates have a stomach, which is simple in shape, as is the human stomach, but they have an enormous and complexly formed cæcum. All the even-toed ungulates, however, whether ruminating or non-ruminating, have, on the contrary, a complex stomach and a simple cæcum. The most complex stomach is that of the deer and hollow-horned ruminants, each of which is provided with four cavities. The freshly cropped food passes down into the first cavity or paunch, whence it goes to the second compartment called, from the character of its lining, the honeycomb bag. Thence it passes upward to the mouth to be chewed once more, whence it descends again to the manyplies, so named from the many folds of membrane within it; and then, last of all, goes to the true digestive chamber of the stomach, called the reed, which opens into the intestine.

Three further points may be noticed, which though in themselves small, yet serve to distinguish the two great groups of ungulates from each other. There is a certain bone of the skull of man and beasts known as the wing-wedge bone, which bone may have a perforation or canal through which a branch of the carotid artery passes. The odd-toed ungulates have this, but the even-toed have it not. The bone of the thigh of man and beasts possesses two bony prominences, or trochanters. The even-toed ungulates like ourselves have these two and no more, but the odd-toed ungulates also possess a third trochanter. The number of bones of the back, together with those of the loins, are seventeen in man. In ungulates they are never less than nineteen, and may be more, as in the horse, where they are twenty-four in number.

Thus the two great groups of hoofed beasts differ as follows:

| Odd-toed. | Even-toed. |
|---|---|
| Functional toes of hind foot, odd. | Even. |
| A simple stomach. | A complex stomach. |
| A complex cæcum. | A simple cæcum. |
| A third trochanter to thigh bone. | Only two trochanters. |
| A canal in the wing-wedge bone. | No such canal. |
| More than nineteen back and trunk bones. | Nineteen back and trunk bones. |

Of the two before-mentioned exceptional animals which have been supposed to be allied to the ungulates, the elephant exhibits all the above given characters of the odd-toed groups save that its thigh bone has but two trochanters. The toes of its hind foot are odd numbered, for there are five of them. Any one observing an elephant when it walks, and noticing the flat sole it applies to the ground, might well suppose it to be plantigrade, and, therefore, quite unlike any other ungulate. But so to think would be a mistake, for it no more applies the sole of its foot to the ground than does a smart lady whose shoe is provided with a fashionable high heel. Beneath the bones of the foot is a great pad, which increases in thickness backward, and thus raises both the heel and the wrist from the ground so much that the elephant is really digitigrade. Nevertheless, the elevation is but slight and enormously less than in any cattle. As a consequence of this latter fact the proportions of the bones of its limbs are more human, and the thigh bone especially is, roughly speaking, so like man's that the finding of such remains might well have given rise to tales about giants.

The hyrax, or coney of Scripture, is a small, short-limbed animal which was originally supposed to be allied to rabbits, rats, and other gnawing animals, or rodents. Cuvier, however, was led by various points in its anatomy, and especially by the structure of its teeth, to associate it with the rhinoceros, in spite of its being entirely clothed with hair, instead of the hairless, folded hides of the great nose-horned beasts of India and South Africa.

Such are the forms of ungulate life now to be found on the surface of this planet, and such the relationships which appear to exist between them. There is a relatively small group of odd-toed beasts containing the three very distinct forms—the horse, the rhinoceros, the tapir—and there is a very large group of even-toed beasts divisible into the two sections, one of swine-like beasts and the other the vast assemblage of ruminants whereof the American bison, or buffalo, is an extreme modification. Besides these there are the two outlying forms—the elephant and the hyrax—between which and the ungulates no connection now exists. But if we go backward in time a very different prospect opens before our eyes. We need only consider the tertiary rocks, the oldest of which is the Eocene formation, and as we recede we shall find a number of gaps to be filled up, the relative proportion of forms to become very different, while relationships, previously unsuspected, between the ungulates as a whole and other orders of beasts suggest themselves to our minds. Thus, the tertiary rocks of France, India, and North America have supplied us with a series of fossil remains which almost entirely bridge over the chasm now existing between the non-ruminating, swine-like beasts, and the ruminants. One of the most interesting, but also one of the most peculiar, of these was also one of the earliest known, having been described and carefully figured by Cuvier from remains found in the gypsum beds of Paris, and named by him *Anoplotherium*. One of its most singular peculiarities consisted in the arrangement and proportion of its teeth, of which it had forty-four, those of each jaw being arranged in an unbroken series all of the same height—an arrangement which is found in no existing animal whatever, except man. Similarly, tertiary deposits in America, India, and Europe seem to show that the distinction we now find between pigs and peccaries had not then arisen. The range of various forms was also

FIG. 55.

THE BRAZILIAN TAPIR.

widely different. The hippopotamus is now strictly confined to Africa, where, besides the common form, there is a pigmy species on the western side of that continent. But in later tertiary times the hippopotamus existed in India and in England; while small species have left their bones in great quantities in the islands of Sicily and Malta. Similarly, as before said, that, in our days, exclusively African form, the giraffe, once existed in Greece, as well as in Persia, India, and China.

As to our type, the bison of North America, remains of a now extinct form have been found, as before said, in Texas, and that has been thought to be the ancestor of the existing species. But if extinct forms of even toed ungulates are numerous, those of the odd-toed ones show that the existing species (horses, asses, zebras, tapirs, and rhinoceroses) are but a poor surviving remnant of a vast quantity of very different species of the group which lived upon the earth's surface during tertiary times. The existing geographical distribution of the tapirs would by itself indicate that they must once have been much more widely distributed than they now are.

Fig. 56.

THE COMMON AFRICAN RHINOCEROS.

In the present day they are found only in South and Central America on the one hand, and Sumatra and Borneo on the other, and all analogy would indicate that they must once have existed in regions intermediate between the New World and the Indian Archipelago. This their fossil remains abundantly prove to have been the case. In Miocene or Pliocene times they existed not only in both China, South Carolina, and California, but also in France, Germany, and even in England. In the same way rhinoceroses which are now found nowhere but in South and Central Africa, India, and the Indian

Archipelago, were once widely distributed, both in Europe and North America; and carcases of a species which was clothed with woolly hair have been found preserved in the frozen ground of Northern Siberia. Another Siberian form, allied to the rhinoceros, had a very large horn supported on a huge prominence, situated further back on the head than is the horn of the existing rhinoceros. A small species found in the Miocene deposits of France, and an allied North American species, are distinguished by the singular fact that they possessed a pair of very small horns placed side by side, instead of one in front of the other, as is the case in all living rhinoceroses which bear two horns. In the United States, during the Miocene period, enormous beasts abounded only inferior in size to the elephant, and which had a transverse pair of large, bony prominences over the nose, each of which probably bore a horny sheath during life. A numerous group of fossil forms existed in Eocene times both in Europe and in America, called lophiodonts. They are characterised by details of tooth structure which it would be out of place here to describe; but the group includes a number of more or less imperfectly known forms, which ranged in size from the bulk of a rhinoceros down to that of a rabbit.

When Cuvier discovered the *Anoplotherium* before noted, he also discovered another interesting, very distinct, fossil animal, which he called *Palæotherium*. It had three digits to each foot, and is interesting as being one of the earliest fossil animals ideally reconstructed by its discoverer. But the fossils which are likely to be the most interesting to our readers are those which relate to that most favourite animal, the horse. There is a kind of wild horse in Central Asia, there are different wild asses in Africa, Syria, Persia, and Thibet; and there are, or till recently were, four wild striped creatures of the horse family in Africa: These were 1. The quagga, with stripes only on the head, neck, and fore part of the body. But a few years ago it was to be met with in great herds, ranging the vast plains which stretch between Cape Colony and the Vaal River. Now, however, it is a question whether any exist save in a few menageries. They have been trained to go in harness, although

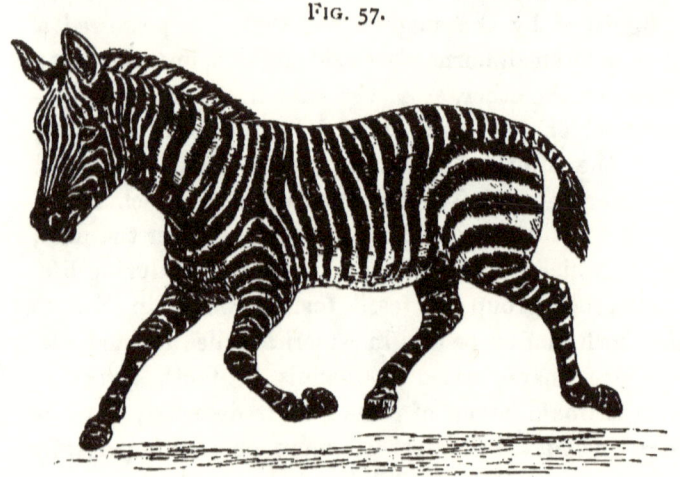

FIG. 57.

THE TRUE OR COMMON ZEBRA.

never really domesticated. 2. The true zebra, which was the one first discovered by Europeans, and which is figured in Buffon's " Natural History," is the most beautifully marked of all, the whole of the body being striped black and white down to the very hoofs. Its natural home was the mountainous country in the vicinity of the Cape of Good Hope, and it has now almost become extinct. 3. The third species is called Burchell's zebra, and used to be highly prized in menageries, as the rarer kind, because of its more distinct home beyond the

Orange River. But it is now the only zebra commonly seen in captivity, and it still roams in herds over the plains north of the river just mentioned. Its numbers are, however, continually diminishing, for zebras are now shot by the natives for their hides, which are very valuable for leather, while their flesh is also much relished. The colour is pale, yellowish brown, except the limbs, which are nearly white. The head, neck, and body, but usually not the limbs, are striped dark brown or black. 4. The fourth and last kind is Grevy's zebra, which inhabits the country south of Abyssinia. It is generally marked like the true zebra, save that the bands are narrower. Such is the distribution of the species of the horse family in the present day.

Wild horses were very common in Europe during what is called the polished-stone period, before the introduction of even bronze weapons. They were domesticated by man before the historical epoch, but the European domestic horse of modern times is in all probability largely, if not mainly, the result of the importation into Europe, through Greece and Italy, of Asiatic horses which were domesticated in times still more ancient. They have been diffused by man almost all over the globe, and in America and Australia, where none existed when those regions were discovered by Europeans, they now roam in great herds. Yet, strange to say, horses were abundant in almost every part of America, from Eschscholtz Bay down to Patagonia, in the most recent geological age, though they had become quite extinct long before the arrival of the Spaniards. A number of fossil remains have been discovered which have been supposed, mainly on account of their tooth structure, to exhibit traces of the real pedigree of the horse. What is certain is that a number of creatures have lived which had toes in various degrees of reduction as to number. Thus a creature known as *Hipparion*, which existed in Europe, Asia, and North America during Pliocene times, instead of having but a single toe to each foot had three; but the two lateral ones were quite minute and rudimentary, and in an Indian variety the lateral toes seem to have disappeared. It was of about the size of a donkey. In the lower Pliocene of North America another species with three digits was also found and was named *Protohippus* by Prof. Leidy. Prof. Cope has suggested that this latter kind was the ancestor of the American horse, and that *Hipparion* was the ancestor of the Old World horse. Should this opinion be verified it would be a complete demonstration that similar forms may have an independent origin, since the horses of America which became extinct closely resembled the horses of the Old World. A fossil beast from a lower formation, that of the Miocene, has been named *Anchitherium*. It was of about the size of a sheep and had three toes to each foot, whereof the lateral ones were less diminutive than in the two kinds last noticed. Sometimes it had in addition a rudimentary middle-foot bone, though with no digit attached to it.

From yet lower rocks, those of the middle and upper Eocene, a still smaller animal has been indicated which has been named *Pachynolophus*, which not only had a large median digit with a smaller one on either side of it, but had also another and still smaller external digit. Lastly, in the lower Eocene remains of yet another beast have been found. It has been named *Phenacodus*, and had five digits to each foot. It has been regarded as the lowest root of that genealogical tree along which the modern horse has been supposed to have been evolved.

We hold our own judgment in suspense as to this question.

Those outlying forms, the hyrax and the elephant, if not actually brought nearer to the existing ungulates by the help of various species now known as extinct, are at least shown to be but some of many others which in different degrees approximate to the true ungulates from various sides. Only two kinds of elephant now exists, and the Asiatic is found nowhere but in the forests of India, Ceylon, Burmah, Cochin China, the Malay peninsular, and Sumatra. The African elephant is confined to the south of the Sahara. Anciently it was fully as much domesticated as is the Indian elephant at present, and bore its part in the armies both of Carthage and of Rome; but now it is only known wild and in menageries. But elephants formerly existed in North America, from Alaska to Mexico, as well as in England, Scotland, and Ireland, Europe and Siberia, where the woolly elephant, or mammoth, has been found frozen, like the woolly rhinoceros. Those elephant-like animals, with simple grinding teeth, the mastodons, ranged over both South and North America as well as India and Europe; whereas other elephantine animals, called dinotheria, which had tusks only in the lower jaw, have been found in Europe and Asia, but not in the New World. Among the most wonderful of all fossil animals ever discovered are certain creatures, as big as elephants, described by Prof. Leidy in 1872. They are among those wonders of an ancient and now extinct world of life which America has made known to us. Their bones were found in the Uinta Mountains, whence they have been named Uinta-beasts, or *Uintatheria*. One of the most curious of their peculiarities was the structure of the head, which bore no less than three pairs of protuberances, each pair being placed side by side. One pair was on the nose, one in front of the eyes, and one on the roof of the skull behind the eyes.

It is America which has alone made us acquainted with four other very exceptional forms, a brief notice of which must conclude this article. One of them was at first thought to be a sort of gigantic llama, and, therefore named *Macrochenia*. And the bones of its neck are formed on the type of those of the camels and llamas. Its limbs, however, are partly like those of the odd-toed and partly like those of the even-toed ungulates. Its grinding teeth resemble those of a rhinoceros, while its cutting teeth have "marks" like those of a horse. The bones of its nose also seem to indicate that it had a short proboscis. It must have been a very curious creature. Another South American extinct mammal was discovered by Darwin near Buenos Ayres and was about as big as a hippopotamus. It seems to have been singularly intermediate between the odd-toed and even-toed groups of ungulates, and has been named *Toxodon*. Yet another South American animal from the same region was rather larger than the giant of the rat and squirrel order, the capybara of the Rio de la Plata, which it resembles in general appearance. Unlike all known ungulates it has collar bones. It has been named *Typotherium*. Numbers of allied kinds have also been discovered. Lastly, in the Eocene rocks of North America the remains of a group of animals have been discovered which have been named by Prof. Marsh *Tillodontia*. Some of these kinds were as large as a tapir.

We said that the discovery of various fossil forms has suggested the existence of relationships and affinities between ungulates and other orders of mammals, which relationships and affinities would not—apart from such evidences—have been suspected. Thus the capybara-like creature noticed last but one, suggests the existence of a distant affinity between ungulates and rodents (hares, rats, squirrels, &c.), while the *Tillodontia*, by the structure of their skull, teeth, and limbs, give rise to the notion that they are related both to the rodents and also to fleshating beasts, or beasts of prey, carnivores. Finally, the dinotherium has a certain affinity to the dugong and manatee—which are marine leg-less creatures, so that if the group of elephants has really any relationship to the ungulates, then these aquatic animals, entirely devoid of hind limbs, must possess such a relationship also.

The results at which we arrive must remain largely speculative, until we have gained much fuller information with respect to forms of life which have for ever passed away. All we can say with certainty at present is, that the odd-toed group and the ruminating and non-ruminating even-toed groups are all very distinct, surviving modified forms of a mass of species which at one time constituted a homogenous group, less diversified in structure than their descendants, and which seems to have been related by affinity to the rodents and possibly also to the carnivorous as well as to the dugong and the manatee. That great group gradually divided itself into two sections, the odd-toed and the even-toed. Of the vast mass of odd-toed forms the immense majority have disappeared, leaving three isolated and very divergent survivors—the tapir, the rhinoceros, and the horse. Of the mass of even-toed forms, a vastly greater number persist, although death and destruction have caused the formation of a wide interruption between swine-like creatures and ruminants. Of the latter, the richest section is that of the hollow-horned ruminants, among which are found those creatures most useful to man, the sheep and the goat, and that animal which was a very providence to the inhabitants of the region which is now the United States when it was first visited by civilised man, namely, the so-called buffalo or American bison.

# VIII

## THE RACOON

THE racoon, absolutely peculiar to North America, may serve as an introduction to some knowledge of beasts of prey and carnivores in general, among which it occupies a somewhat exceptional position. It is a stoutly built quadruped, although its coat of long coarse hair makes it look yet stouter than it really is. It is about the size of a badger, and has a sharp-pointed muzzle, rather short ears, and a moderately long, bushy, but cylindrically-shaped tail, marked with black and white rings. The general colour of its hair is greyish brown, and there is a light-coloured patch over either eye and on the side of the muzzle. The limbs are of medium length, and each paw has five toes. Those of the forepaws can be stretched wide apart and all the digits have arched and pointed claws which are not retractile, like those of a cat. When standing, the soles of the feet are wholly applied to the ground, so that the animal is what is called plantigrade, although in walking the heel is somewhat raised. Its grinding or molar teeth are mostly broad and rather flat, with moderate prominences, and no remarkably sharp blades adapted for cutting flesh. Its name of racoon, familiarly abbreviated into coon, is a corruption of its Indian designation arathcone. It can make a good fight, an old coon being a good match for an average dog. Though very sly, racoons are caught in traps. They are not swift runners, and if pursued take to a tree. The racoon, though capable of being made a pet of, cannot be let loose with impunity, on account of its great curiosity, which leads

FIG. 58.

THE RACOON.

it to find its way into the house and examine everything. It is very fond of sweets, and will remove covers from dishes, corks from bottles, and soon learn to unlatch doors. Its natural range extends all over the United States, both north to Alaska, and south to Costa Rica, attaining its largest size in the South.

It stirs abroad but very little by day, and only when the weather is dull and cloudy. No North American animals are more strictly nocturnal, except bats and flying squirrels. And it not only sleeps by day, but also during the winter. Yet it makes no specially comfortable nest wherein to repose, but only coils itself up in a hollow tree or, by preference, in some dead branch; for it chooses, no doubt for greater security, an elevated position.

Racoons are carnivorous animals, and sometimes they eat poultry, but mice, small birds, eggs, insects, fruits, nuts, maize, frogs, crustaceous molluscs, and fish, are all welcome food to them. They are very expert in breaking down stacks of corn and stripping the husks from the corn, using their paws like hands. They swim well and will cross rivers without hesitation, but they cannot dive and pursue fish under water as otters do. They readily, however, obtain crawfish and mussels. They like to play in shallow water and overturn stones in search of crawfish, and they have a singular habit of washing their food in water before eating it. There is a Southern species, called the crab-eating racoon, but that term could also be supplied to the Northern kind. The Southern one has, as might be expected, shorter fur, and it has also stronger teeth, but otherwise is very like the Northern kind, both in structure and habits. It is to be found all over South America as far south as the Rio Negro.

There are two beasts, closely allied to the racoon, but more slender in build and with longer tails, found in some parts of the United States and North Mexico. One has been captured in Ohio, and in Oregon north-west of Jacksonville. Catamiztli was a name applied to this kind in Mexico, and it is also called cacomistle and the cat-squirrel by the Texans. Its real relationship to the racoon was long unsuspected, as it was taken to be one form of the very different group of civets. The Central American species is called Sumichrast's cat-squirrel. These two beasts go more on the tips of their toes than

Fig. 59.

THE COATI-MUNDI.

do the racoons. They are readily tamed, and are made pets of by the miners of California. They dwell in woods, and make a moss-lined nest in a hollow tree, and are often betrayed by chips of wood which they will gnaw off round the mouth of the hole they inhabit. Their food consists of small kinds of beasts

## THE RACOON

and also insects. They are useful for destroying mice and rats, but are very destructive to poultry, and are naturally bold, and will fight furiously with claws and teeth. They prefer to inhabit woods traversed by water-courses. Two more species of animals, also entirely confined to the New World, are known as coati-mundis,

FIG. 60.

THE KINKAJOU.

or coatis. One of these is confined to Mexico and Central America, and the other to South America, from Surinam to Paraguay. They are not so stoutly built as are the racoons, and have longer and more slender and tapering tails, but their main peculiarity consists in the possession of a very elongated and mobile snout or short proboscis. The coatis live mainly in trees, going about in troops of from eight to twenty individuals. They are also, like racoons, indiscriminate feeders, eating fruit and insects as well as birds and eggs.

South and Central America produce another kindred animal, though very distinct in aspect and organisation. This is the kinkajou (Fig. 60), a strictly forest creature, found in the warmer parts of South and Central America. It has a long body, but short limbs, which are well fitted for clinging to the trunks and branches of trees by the very strong and sharp claws with which all the toes are provided. But it is still better fitted for arboreal life by means of its tail, which is very long, and also strongly prehensile, like the tails of so many American monkeys. The kinkajou is of about the size of a rather small cat, and is clothed with short, dense fur of a uniform pale yellowish-brown colour. It has a broad, round head, with very short ears and an extremely long and very extensile tongue, which is, no doubt, of much use to it in eating honey, of which it is very fond, although it will also devour eggs and small birds and beasts. It is a nocturnal animal of rather a gentle disposition, and it is easily tamed. In captivity it will live on oranges and bananas, which it eats greedily. It is not uncommonly found in holes of trees, where it lies concealed by day, issuing forth at night in pursuit of prey. Its woolly fur is much valued, and the skins are brought to market. Dampier, in his "Voyages," says : " The flesh is good, sweet, and wholesome meat. We skin and roast it, and then we call it pig, and I think it eats as well." There is yet another exclusively American animal—another sort of cacomistle, or cat-squirrel, about which a few words must here be said. There are two kinds, one named after Mr. Gabb and the other after Mr. Allen, and coming from Costa Riça and Ecuador respectively. Of one kind, only the skull is known, but the other has a very long body, and tall and short legs, and from its form and coloration it is so like a kinkajou that the Indians who accompanied its discoverer, the collector, Mr. Buckley, actually mistook it for that animal, although the snout is longer relatively and the head less rounded. An English naturalist, Mr. Oldfield Thomas, who has figured it, considers it to be a real case of mimicry. He observes, however, that it is very difficult to understand what benefit to the creature it can be to be mistaken for the kinkajou, though a knowledge of its habits when gained may explain this.

Thus, the racoon is a common familiar representative of a small group of quadrupeds almost entirely confined to America, and which are beasts of prey indeed, but, as it were, of a mild and moderate kind, forming a sort of intermediate group between the more predaceous families, such as those which respectively contain the tiger, the wolf, the grizzly bear, and the weasel. The racoon tribe, or racoon family, has, however, a representative in the Old World, which is found in the south-eastern Himalayas at from 7000 to 12,000 feet above the sea. It is known as the panda, is rather larger than a cat, and has semi-retractile claws. Its fur is of a rich reddish-brown colour, with the under part of the body the darkest, and its long tail has dark annulations. It lives mainly on fruit and other vegetable substances, such as acorns, sprouts of bamboo, roots, &c., and rarely eats flesh, although it is said to take insects. It frequents the woods of rocky regions. It is not a strictly nocturnal creature, although it sleeps much by day, coiled up like a cat, roaming abroad each morning and evening. None of its senses are acute, and it is easily caught, being neither cunning nor ferocious. It drinks by inserting its lips into the fluid. The panda is easily tamed, but cannot endure either much heat or much cold. The young are born in a very helpless state, and long remain hidden in the nest placed in a hollow tree or in the fissure of some rock. It is a very interesting fact that a creature of this kind lived in England in the latter tertiary times, and thus the present scattered distribution of the racoon family is bridged over by the help of fossil remains, just as is the present scattered distribution of the family of tapirs among the hoofed beasts.

The racoon tribe has been often supposed to be more nearly allied to the bears than to any other group of flesh-eating beasts, a relationship we much doubt. Whatever may be the value of this supposition, however, the Rev. Père David discovered in 1869, in the almost inaccessible mountains of Moupin, in Thibet, a creature which, though it has a very short tail, is in some respects like the panda, though in others like a bear. It is said to feed principally on vegetable substances, such as bamboos and the roots of various plants.

We may now pass on to consider the bear tribe. There are ten kinds of true bears, which range from the Arctic regions southward to Africa, north of the Sahara, the Indian Archipelago, and Chili. No species, however, is common to the Old World and the New. None is found in central and southern Africa, none in Australia, and only one in South America. The bears are animals of considerable size, and among them are found the giants of the carnivorous order. They walk on the naked soles of their feet, have very short tails, moderately short, erect ears, small eyes, and fur which is generally long and shaggy, with claws which are long, strong, and non-retractile. The great white or polar bear of the Arctic regions is of the same colour all the year round. It lives on animal food, and very largely on fish; yet in the summer time it will eat a large quantity of grass. The commonest and best known bears are the brown bear of Europe and northern Asia and the grizzly bear of North America—forms which some naturalists consider as merely varieties of the same species. Anyhow they are both very formidable animals. In cold regions the brown bear hibernates, and the varieties which inhabit warmer climes are smaller in size than the northern forms. They once inhabited England, are still found in the Pyrenees, and are numerous in parts of Russia and also in Norway and Hungary. In the Himalayas, where they are now very numerous, they live at high elevations, and they come out of their winter sleep about March or April, when they feed largely on the bulbs of plants. They are very fond of succulent sweet fruits, but are also often carnivorous, killing sheep or goats, or even cattle, and an instance has been recorded of a large bear killing two small ones and eating portions of them. Bears generally walk slowly, but they can run pretty quickly in a clumsy gallop. The young are born very small, scarcely larger than a good-sized rat; they are hairless, and remain blind for four weeks. Cubs of two different years are often found with the mother at the same time, and all remain with her till nearly three years old. Bears, as every one knows, are easily tamed, and they are also long-lived. One of them, maintained by the State of Berne in Switzerland, lived for forty-seven years, and a female thirty-one years old bore young. Neither the brown bear's sight nor its hearing is acute, but it has a delicate sense of smell. In the Himalayas and also in Persia and China a black bear is to be met with, and other species of black bears are found in Japan and North America. They are forest animals and eat fruit largely, as also maize and nuts. At the same time the black bear is the most carnivorous of the Indian bears. It kills sheep, goats, deer, and even cattle and ponies, but it occasionally feeds on carrion. Many natives are killed or severely injured by it. Its senses are more acute than are those of the brown bear, and it is an excellent swimmer.

The single kind found in South America is called the spectacled bear, from the markings round its eyes. It inhabits the Peruvian Andes. The Malay bear is a

FIG. 61.

THE SLOTH BEAR.

small species with a black coat, strongly curved claws, small rounded ears, and a very long tongue. It is found from Burmah to Borneo, where it dwells in forests, being essentially frugivorous, though sometimes eating small beasts and birds. The most curious of all the bears is the sloth bear, an animal clothed with very long and coarse black hair, with an elongated greyish snout, a white horseshoe mark on the chest, and very long and curved white claws. Its eccentric appearance is due not only to its long, shaggy coat, but also to its peculiarly shaped head, long, mobile snout, and short hind legs, and likewise to its queer antics. In spite of the number which sportsmen have destroyed, this is still one of the commonest wild animals of India, where it ranges from the Himalayas to Cape Comorin and Ceylon. The following facts concerning this interesting animal are recorded by Mr. Blanford, F.R.S.: Its food consists almost entirely of fruits and insects, beetles and their grubs, the honey and young of bees, and the combs of white ants. Its powers of suction and of propelling wind from its mouth are very great, and it is thus enabled to procure its white ant food with ease. On arriving at an anthill the bear scratches away with his forefeet until he reaches the large combs at the bottom of the galleries. He then with violent puffs dissipates the dust and crumbled particles of the nest, and sucks out the inhabitants of the comb by such forcible inhalations as to be heard at two hundred yards distance or more. In southern India these bears are fond of the fermented juice of the wild date palm, and climb trees to get at the pots in which it is collected. The animals are said at times to get drunk with palm juice. They are very fond, too, of sugar-cane, and do much damage to the crops. Bears generally have a habit of sucking their paws and of making at the same time a peculiar humming sound, and this is especially the case with the sloth bear. Except as regards this puffing and humming, the sloth bears are usually silent animals and have no "call" for each other. When surprised or disturbed, however, and especially when wounded, they become very noisy, uttering a series of loud, guttural sounds, and when mortally wounded emitting peculiar wailing cries. As a rule the sloth bear is a timid animal, but occasionally it attacks man savagely, using both its claws and teeth, and clawing the head and face of its victim. Sometimes, especially when surprised suddenly and attempting to escape, a bear merely knocks a man down with a blow of its claws, often, however, inflicting severe wounds; but in other cases it holds him with its claws and bites him savagely, not leaving him until after he ceases to struggle, and sometimes an onslaught appears quite unprovoked. The pairing time is mostly in June, and the young are generally born in December or January. When about two or three months old the mother takes her pair of cubs with her, carrying them on her back, where they cling to her long hair. They sometimes ride thus until of tolerable size, and one cub may sometimes be seen following its mother, while the other is carried. They take two or three years to reach maturity, and have been known to live in captivity for forty years. They are easily tamed when caught young, and, although fretful and querulous at times, are generally playful, amusing, good-tempered, and much attached to their masters.

The family of bears agrees with the group of smaller animals, whereof the racoon is the type, in having grinding teeth in considerable number, but more or less blunt, and but very slightly adapted for cutting flesh. The claws are long and powerful, but never more than semi-retractile—as in the panda. With these animals, then, it will be profitable to contrast those creatures, the bodies of which are constructed in the most perfect manner known to us for a predaceous life—those which are so specially modified in tooth and claw as to fit them, beyond all other animals, for carnage and destruction. Such, especially carnivorous animals, are the lions and tigers of the Old World and the jaguar and puma of the New, and a perfect type of all such animals is presented to us by the structure of the cat, for the whole cat tribe are formed almost in entirely the same manner. The reader is referred to a work of ours, entitled "The Cat," published by John Murray of London and Charles Scribner's Sons, New York. Therein the anatomy, physiology, development, and general natural history of the cat and cat tribe are set forth in detail. The cat's teeth are admirably adapted for cutting flesh, two opposite grinders especially, one in the upper and one in the lower jaw, having sharp cutting blades, which play one against the other like the blades of a pair of scissors. On this account these two teeth are known as the "sectorial" teeth. The small number of the teeth is also a noteworthy character. Behind the large eye teeth, or canines, there are in the lower jaw but three teeth on each side. Of these, the first is but a small one, while the third and largest is the lower sectorial tooth. In the upper jaw, the first tooth behind the canine is exceedingly small, and in some kinds of the cat tribe is wanting altogether. The third tooth is much the largest, and, with its sharp cutting edges, is the upper sectorial. Behind this is a very minute tooth, which has no cutting edges, but is like each of the grinders of the bear and racoon family, a tubercular tooth Except in the hunting leopard, or cheetah, all the claws are completely retractile.

There are more than forty different species of the cat tribe. As was long ago remarked by Buffon, the great cats of the Old and New World are markedly distinct. The lion, tiger, leopard, ounce, clouded tiger, caracal, and cheetah, with a variety of smaller cats, are all inhabitants of the Old World only. The puma, jaguar, ocelot, jaguarondi, eyra, collocolli, the pampas cat, and one or two others, are exclusively inhabitants of the New World. It is only among the lynxes that we find a form common to both sides of the Atlantic, the Canadian and north European lynxes being probably but varieties of one species. America is not so rich in species of the cat tribe as is the Old World, nor do its largest kinds, the puma and jaguar, equal the largest kinds of Asia and Africa. Strange to say, the West Indian Islands, although some of them, as Cuba and Hayti, seem admirably suited to shelter and support species of the cat family, are entirely destitute of them. The same is the case with the great island of Madagascar (in spite of its forests, with their numerous animal population), and also with New Guinea, New Zealand, and the whole of Australia. America, north of Arkansas and Louisiana, has the lynx and the puma. Europe has two species of lynx and the wild cat. It might be supposed that the domestic cat is simply the wild cat tamed, but it is probably a descendant of the Egyptian cat, which was domesticated in very ancient times. Egypt, as the granary of the ancient world, might well have been the country in which it was originally tamed. It was certainly domesticated there 1300 years before Christ, and there is a painting in the Egyptian Gallery of the British Museum of a tabby cat which seems to be aiding a man in the capture of birds. The Goddess Pasht, or Bubastis, the Goddess of Cats, was, under the Roman Empire, represented with a cat's head, and a temple at Beni Hassan dedicated to her is as old as 1500 B.C., while behind it are pits containing a multitude of cat mummies. The cat was an emblem of the sun to the Egyptians, and its eyes were supposed to vary with the course of that luminary, and it is a fact that the eyes of at least some cats do really change colour. Herodotus relates extraordinary stories of the veneration in which this animal was held by the Egyptians. He tells us that when a cat died a natural death in any house, the inhabitants of it would shave off their eyebrows, and that when a fire occurred their first care was rather to save their cats than to extinguish the conflagration. The domestic cat was a very precious animal in Western Europe in the Middle Ages, as is shown by the heavy fines imposed on those who should destroy one in Wales, Switzerland, or Saxony. As compensation, a payment was required of as much wheat as was needed to form a pile sufficient to cover over the body of the animal to the tip of its tail, the tail being held up vertically, with the cat's muzzle resting on the ground.

The wild cat is still to be found in Southern Russia and the adjacent parts of Asia, Turkey, Greece, Hungary, Germany, Dalmatia, Spain, Switzerland, and, though now very rare there, France. Thanks to the destruction of forests in England, and the over-zeal of gamekeepers, the wild cat is now extinct in England, and perhaps in Wales also, although it was to be found in Wales thirty years ago, and in England sixty years ago. In Scotland it is still far from uncommon, especially in Inverness, Ross-shire, and on the West Coast of the Highlands. It is also found in Skye, but seems never to have existed in Ireland.

All the various kinds of cats, from the lion downward, live naturally on warm-blooded animals which they have themselves killed. The only exception is the Indian fishing cat, which, besides fish, will eat fresh-water mollusks. The different species are not only very uniform in structure, but the uniformity of the colour is also remarkable. Some reddish, or yellowish shades more or less modified by grey or brown, may be said to be their ground tint, and this is generally marked with spots, or stripes of black, while the under parts of the body are whitish. A few species, however, as the lion, the puma, the jaguarondi, and eyra, are uniform in colour. Cats never hunt in packs as dogs and wolves do, and rarely pursue their prey in open ground, but spring upon it from some hiding-place. They are mostly nocturnal, and the greater number, especially of the smaller kinds, habitually live in trees. The lion, the male of which distinguished from all other cats by its mane, is now found only in Africa, Mesopotamia, Persia, and North-Western India, although formerly it existed all over India and in South-Western Europe, the camels of the army of Xerxes having been attacked by lions in Macedonia. It frequents sandy plains and rocky places, and is most active at night. As to the African lion, Drummond tells us: " I once had the pleasure of, unobserved myself, watching a lion family feeding. I was encamped in Zululand, and toward evening, walking out about half a mile from camp, I saw a herd of zebra galloping across me, and when they were nearly two hundred yards off I saw a yellow body flash toward the leader, and saw him fall beneath the lion's weight. There was a tall tree about sixty yards from the place, and, anxious to see what went on, I stalked up to it while the lion was still too much occupied to look about him, and climbed up. He had by this time quite killed the beautifully striped animal, but instead of proceeding to eat it, he got up and roared vigorously, until there was an answer, and in a few minutes a lioness, accompanied by four whelps, came trotting up in the same direction as the zebra, which, no doubt, she had been driving toward her husband. They formed a fine picture as they all stood round the carcass, the whelps tearing it and biting it, but unable to get through the tough skin. Then the lion lay down, and the lioness, driving her offspring before her, did the same four or five yards off, upon which he got up, and, commencing to eat, had soon finished a hind leg, retiring a few yards on one side as soon as he had done so. The lioness came up next and tore the carcass to shreds, bolting huge mouthfuls, but not objecting to the whelps eating as much as they could find." Every one who has heard it is deeply impressed with the lion's wonderful voice when he emits loud, deep-toned roars in quick succession, which get louder and louder, and are then succeeded by muffled sounds like distant thunder.

The tiger is an Asiatic animal exclusively, and ranges, in suitable situations, from the Amoor to the Island of Bali, and from Turkish Georgia to the Island of Saghalin, but does not exist in Ceylon. In spite of the great destruction of tigers in India, they still live, according to Mr. Blanford, wherever large tracts of forests and grass-jungle exist, and they are specially common in the forests at the base of the Himalaya. Tigers at least occasionally accompany the tigress and her cubs, for these animals, like lions, are monogamous. The young remain with the mothers until nearly or quite full grown. By day the tiger takes up its abode in deep shade, especially in the hot season, and generally near water. They swim well and will even cross arms of the sea, but very rarely ascend trees. Tigers spring much less than is popularly supposed, and rarely move both their hind legs off the ground. They roar a good deal less than lions do, although their call is very similar. Mr. Blanford says: "When hit by a bullet a tiger generally roars, but tigresses, generally, or at all events very often, do not. I have on three occasions at least known a tigress receive a mortal wound and pass on without making a sound." The ordinary food of tigers consists of pigs, deer, antelopes, and, strange to say, porcupines, which one would think would be rather awkward mouthfuls. They also sometimes kill and eat bears and young gaurs and buffaloes, although such wild cattle, if adult, are more than a match for the tiger. When hard pressed during inundations they will eat fish, tortoises, lizards, frogs, and even locusts. They kill great numbers of domestic animals, and sometimes live entirely on cattle, and they have a distinct preference for beef over mutton. The tiger appears ordinarily to kill cattle by clutching the forequarters with its paws and then seizing the throat in his jaws from underneath and forcing it upward and backward until the neck is dislocated. The enormous muscular power of the tiger is shown by the way in which it can transport large carcases of oxen over rough ground, sometimes lifting the body completely off the surface. A very hungry one will devour two hindquarters in one night, but generally remains three or four days near the carcass, feeding at intervals. A tigress with cubs is often very destructive, partly, it is said, in order to teach the young tigers to kill their own prey. Though they usually do so kill, they do not disdain carrion. Cases are even recorded of a shot tiger being devoured by another of its own species.

The ordinary cattle-eating tiger is a great coward in the presence of man, and often allows himself to be pelted off. The man-eating tigers are those which have got fat and heavy, or, being disabled from age or injury, find man an easy prey; and when once they have got over their innate fear of the human species such a tiger may become a fearful scourge. Thus, in Lower Bengal alone 4218 persons were killed by them between 1860 and 1866. In Bengal and Upper India tigers are hunted on elephants, the sportsmen shooting from howdahs. In Central and Southern India tiger shooting is chiefly attempted in the hot season, and the tiger is either driven by beaters past a tree on which the sportsman sits, or followed up, either on an elephant or on foot. Occasionally, especially when a tiger has been wounded, a herd of buffaloes are employed to drive him out of the cover, which they do very effectually, charging him in a body if he does not retreat. Tigers captured young, are easily tamed, and many of the adult animals in menageries are perfectly good-tempered, and fond of being noticed and caressed by those whom they know. They have repeatedly bred in confinement, although not so freely as lions, and the cubs more rarely thrive.

That beautifully spotted animal, the leopard, or panther, is a cat which has a very wide range, namely, from Algeria to Cape Colony in Africa, and in Asia from Palestine, China and Japan, to Ceylon, Java, Sumatra, and Borneo. In early times it also existed, as we know from fossils, in Great Britain, France, Germany, and Spain. Perfectly black leopards, which, however, in certain lights, show the characteristic markings of the fur, are not uncommon. The habits of leopards differ materially from those of tigers. The leopard is much more lithe and active, climbs trees readily, and makes immense bounds clear off the ground. It is as bloodthirsty and ferocious as any of the cat tribe; it is bolder than the tiger, and not unfrequently attacks our own race. Instances have been known of one becoming a regular man-eater, and such a leopard is said to have killed in two years no less than two hundred human beings. Leopards are very fond of eating dogs and jackals, and are terrible foes to monkeys. The ounce is a lighter-coloured leopard, with longer fur, which inhabits the highlands of Central Asia, ascending to altitudes of from 9000 to 18,000 feet above the sea. The American so-called lion, or puma, ranges from Canada to Patagonia, and reached at one time from the Atlantic to the Pacific. It is still common in the dense forests which clothe the mountains of Central America. The most powerful of the American cat tribe, the jaguar, only extends from Texas southward nearly to Patagonia.

The cats exhibit to us a structure of body specially modified for a predaceous existence. Nevertheless, certain extinct animals of the group had attained a more special and extreme organisation of the kind than is to be found in any existing species. These were the sabre-toothed tigers, remains of which have been discovered in different tertiary rocks in India, Europe (including England), and both North and South America. They had enormous canine, or eye teeth, the tusks of the upper jaw attaining a length of seven inches in one South American form which was about the size of a tiger. Also, the blades of these teeth were much flattened from within outward, their sharp, cutting edges being serrated like a small saw—a character but feebly developed in any of the large living cats. Moreover, the lower jaw was sometimes much broadened from above downward, the better to protect these enormously developed teeth, which in some species were so large that the jaws could not be opened beyond them, so as to allow them to be used for biting. They could therefore, only be made use of as daggers, the animal striking with them while its mouth was closed.

Naturalists are now agreed that the group which includes the civets, the genets, and the mongooses is one nearly allied to the cat tribe. It is a large and varied group which includes many species, but not a single one of them is to be found, except in confinement, on the American side of the Atlantic. They are creatures

# THE RACOON

with longer and more pointed muzzles than the cats, standing lower on the limbs, and with a second blunt or tubercular molar, or grinding tooth, on either side of the

Fig. 62.

THE POIANA.

upper jaw, and claws less retractile than those of the cats. The civets are handsomely marked, striped and spotted animals, as are also the genets and three allied forms called linsangs, which come from India and the Malay Archipelago. The civets, about the size of a large fox, are the largest animals of the whole group. They are found in Africa and India, and produce that substance, smelling so powerfully of musk, which has been an object of commerce for centuries.

The Indian civet extends from Southern China to the Malay peninsula. It lives generally in a solitary fashion in woods or thick grass during the day, whence it comes forth at night, often entering houses. It is very destructive to poultry, and kills any birds or small beasts it meets with, but it will also feed on snakes, frogs, insects, eggs, fruit, and roots. The genets are but of the size of small cats, and have long and slender bodies: all of them are exclusively African except the common genet, which is a native also of the south of France, Spain, and South-Western Asia. A handsome species, which has been named *Poiana* (Fig. 62), comes from Sierra Leone; and there is a small group of allied Indian forms known as palm-civets and toddy cats, which also got the name of paradoxures from F. Cuvier on account of a peculiarly curled condition of the tail. The long tails of these animals are not truly prehensile, but they can coil them to some extent, and in caged specimens the coiled condition not unfrequently becomes confirmed and permanent. These animals are found from China and Nepal, to Ceylon, Java, Borneo, and the Philippine Islands. Mr. Blanford tells us that the common species, the Indian palm-civet, is a familiar animal enough in most parts of Hindostan, although rarely seen by daytime, as it is thoroughly nocturnal. It generally passes the day in trees, either coiled up in the branches or in a hole in the trunk. Cocoanut palms and mango groves are favourite resorts. It also not unfrequently takes up its abode in the thatched roofs of houses. It feeds much as the civet does, and when taken

## THE RACOON

young is easily tamed. A nearly related species from Burmah and the Indian Archipelago has been named *Arctogale*.

Fig. 63.

THE ARCTOGALE.

A very exceptional form of allied animal is found in the Malay Archipelago—the bear-cat, or binturong. Its head and body are about two and a half feet long, and its tail is also upward of two feet. It has tufted ears, short limbs, and very long and harsh fur. It is a good example of the small value which characters drawn from the teeth only, possess as evidences of essential affinity between different animals—for the bear-cat's teeth are very small, and separated from one another by very distinct intervals. It is a strictly arboreal animal

Fig. 64.

THE BINTURONG.

with a slightly prehensile tail, and is very slow and cautious in its movements, and it is a creature of the civet tribe, specially modified for tree life. But another still more exceptional form is one specially modified for aquatic life, although it is also able to climb trees. It is known as the *Cynogale*, and has short toes which are slightly webbed, a long head, with a broad, flattened muzzle, very long whiskers, and short ears and a short tail. It is also an inhabitant of the Malay Archipelago, and lives on fish, crabs, small beasts, and birds, and also, it is said, on fruit.

Of the mongooses there are some twenty-one different species, the majority of which are Asiatic, and they have all relatively longer and more slender bodies, shorter limbs, and more pointed muzzles than the civets. The mongoose lives principally upon rats, mice, snakes, lizards, eggs, insects, and any small birds it can catch. It is an excellent ratter, and is said to be able to kill a dozen full-grown rats in a minute and a half. Its introduction into Jamaica is said to have resulted in a saving of more than £100,000 a year in protecting the sugar canes from rats. Much has been written about the combats between this animal and venomous snakes, and the immunity it is said to enjoy from the effects of the serpent's bite. Mr. Blanford, whose experience is so great, observes: "The prevalent belief throughout Oriental countries is, that the mongoose, when bitten, seeks for an antidote, a herb or root known in India as *manguswail*. It is scarcely necessary to say that the story is destitute of foundation. There is, however, another view, supported by some evidence, that the mongoose is less susceptible to snake poison than other animals. I have not seen many combats, but so far as I can judge the mongoose usually escapes being bitten by his wonderful activity. He appears to wait until the snake makes a dart at him, and then suddenly pounces on the reptile's head and crunches it to pieces. I have seen a mongoose eat up the head and poison glands of a large cobra, so the poison must be harmless to the mucous membrane of the former animal. When excited, the mongoose erects its long, stiff hair, and it must be very difficult for a snake to drive its fangs through this, and through the thick skin which all kinds possess. In all probability a mongoose is very rarely scratched by the fangs, and if he is very little poison can be injected. It has been repeatedly proved by experiment that a mongoose can be killed like any other animal if properly bitten by a venomous snake, though even in this case the effects appear to be produced after a longer period

Fig. 65.

THE FOUSSA.

than with other beasts of the same size." There are some small allied forms in Madagascar, where alone is also to be found the most catlike of the whole group— the foussa. This animal was described and named many years ago in the first volume of "The Transactions of the Zoölogical Society of London," and has been recently imported into that society's gardens.

To the tribe of civets, genets and mongooses, succeeds another small group of exclusively Old World mammals, which are different indeed in size and form from the last-named animals, and yet in essential structure are closely allied to them. The group referred to is that of the hyenas, whereof there are three species—one common to Northern Africa and Southern Asia, and the other two confined to Africa south of the Sahara. No predaceous animals have teeth so powerfully formed for crushing bones as have the hyenas, and they will eat up such as have had the flesh picked off by vultures and jackals. But although the main food of the hyena consists of the bodies of animals which it finds already killed, it will occasionally carry off to its den living sheep, or goats or dogs, and there devour them. Mr. Blanford tells us that in India the hyena is universally despised for its cowardice, and that in spite of its powerful teeth it rarely attempts to defend itself. It is occasionally ridden down and speared, but unless the ground is peculiarly favourable for horses, it will give a good run before being killed, not on account of its speed, for it is easily caught by a good horse, but from the way it turns and doubles. An instance is related in which a hyena, after being slightly wounded by a spear, was pursued by a game old Arab horse who had lost his rider, and who attempted to seize the hyena with his teeth and to strike him with his fore foot, an attack which the hunted animal only acknowledged by tucking its tail tightly between its legs. Hyenas are easily tamed if captured young, and become very docile and greatly attached to their masters. In ancient times hyenas were common not only on the continent of Europe but also in England.

The binturong has already shown us how the teeth of one species of a group may vary by defect from those of its congeners. This fact is made yet more evident by a South African animal named aard (or earth) wolf by the Dutch colonists. Save for its greater slenderness, this animal has the general form of a hyena, with very erect and pointed ears and a well-developed mane along the middle line of the neck and back. Its mane can be erected at will, and gives the animal a formidable look. But it is a mere game of brag, for its teeth, except the canines, are quite rudimentary and insignificant. Observations made upon specimens in menageries show its harmlessness, and that it has neither the inclination nor the power to feed upon living beasts and birds. It feeds only upon decomposing animal substances and upon grubs and white ants, and is a nocturnal, burrowing beast. Thus, the civet tribe and the hyenas are, as it were, zoölogical cousins to those most perfect of predaceous beasts, the cats.

The bears, on the other hand, have also been supposed to have for their zoölogical cousins creatures which are mostly small in size, though many of them are great in value, since among them are included the sable, the ermine, and the mink, and other kinds zealously hunted for their valuable fur. For ample details the reader is referred to a work compiled in 1877 by Dr. Elliott Coues on "Fur-bearing Animals," printed at the Government Office at Washington. They may be spoken of as of the weasel tribe, as we speak of the cat tribe and the civet tribe. The largest of the weasel tribe, and the most bear-like in aspect is the glutton, or wolverine, which has a heavy body supported on thick-set, rather low legs, with nearly plantigrade feet, and long, curved claws. It ranges over the north of both hemispheres, descending in America down to the borders of Arizona and New Mexico. Its food consists of hares, foxes, beavers, squirrels, grouse, and reindeer, and it is said to attack even horses and cows. It has a curious propensity to steal and hide things, even objects which can be of no possible use to it. Dr. E. Coues relates a singular instance of this habit. A hunter and his family having left their lodge unguarded during their absence, on their return found it completely gutted. The walls were there, but nothing else. Blankets, guns, kettles, axes, cans, knives, and all other paraphernalia of a trapper's tent had vanished, and the tracks left by the beast showed who had been the thief. The family set to work, and by carefully following up all his paths recovered, with some trifling exceptions, the whole of the lost property. The common badger is another of the weasel tribe of considerable size, being from two and a half to three feet long, and

FIG. 66.

THE AMERICAN BADGER.

standing about one foot high at the shoulder. There are some half-dozen species of true badgers, all peculiar to Europe and Asia, including England and Japan.

Another kind is the hog-badger, which is found nowhere but in Central and Southern Asia. A small, burrowing kind, called the teledu, is found in the mountains of Java, while a form quite different from all these—the American badger or braro—is widely spread over North America. It is, however, most rarely seen, since it not only burrows, but lives almost as much under ground as does a mole, while in colder latitudes it hibernates during a considerable portion of the year. Dr. Coues says that he has travelled for days and weeks in regions where these badgers abounded, and where their innumerable burrows offered the principal obstacle to progress on horseback or by wheeled conveyance, yet he never saw more than five, and they were in sight but a few moments as they hurried to the nearest hole. They prey upon other and much smaller burrowing beasts belonging to the rat and field-mouse order, and also on insects, snails, and birds' eggs. They are very fond of the stores of wild bees, the honey, wax, and grubs being alike devoured. But the creature of the weasel tribe most notoriously fond of the honey is the ratel of Africa and India, a beast very much of the build of a badger, with a head and body a little over two and a half feet long and a tail half a foot long. It is found all over Hindostan, but not in Ceylon, and is strictly nocturnal, remaining in its burrow by day. Though it eats honey when it can get it, rats, birds, frogs, and insects are eaten also. It is, by the natives of India, suspected of digging into graves to eat the bodies of the dead, but there is no foundation for such a suspicion.

The martens, sables, polecats, stoats, ermines, and weasels form a very natural assemblage of long-bodied, short-limbed, small beasts of prey, exceedingly bloodthirsty in their habits. The martens are only found in the northern hemisphere, where they range through the greater part of both the Old World and the New. Their sectorial teeth are very efficient flesh cutters, and there is but one tubercular tooth behind it in either jaw. They live in trees and climb with great facility. The sable of the Old World is found chiefly in Eastern Siberia. The New World sable is extensively distributed in North America from Newfoundland to Colorado. In spite of the persistent and uninterrupted destruction to which it

THE RACOON 241

is subjected, Dr. Coues tells us that up to 1877 it did not appear to have materially diminished in numbers in unsettled parts of the country, and this, although the

FIG. 67.

THE SKUNK.

annual imports into Great Britain have exceeded 100,000. That beautiful little animal, the ermine, ranges over Northern Asia, Europe, and America. The tip of its tail remains black when all the rest of its fur

Q

turns white. The utility of this persistence to the animal itself is problematical. It is certainly the cause of its persistent pursuit by man. One most curious member of the weasel tribe is the skunk (Fig. 67), an animal conspicuously marked with black and white, and with a long tail handsomely clothed with abundant fine hair. The common skunk ranges from Hudson's Bay to Guatemala. It preys on small beasts, birds, and reptiles, but especially on insects and mice. As to this animal A. R. Wallace tells us that while staying a few days in July, 1887, at the Summit Hotel on the Central Pacific Railway, he strolled out one evening after dinner, and on the road, not fifty yards from the house, he saw a pretty little white and black animal with a bushy tail coming toward him. As it came on at a slow pace and without any fear, although it evidently saw him, he thought at first that it must be some tame creature, when it suddenly occurred to him that it was a skunk. It came on till within five or six yards of him, then quietly climbed over a dwarf wall and disappeared under a small outhouse, in search of chickens, as the landlord afterwards told him. As is well known, the skunk possesses the power of ejecting a most offensive secretion, and this effectually protects it from attack. The odour of the substance is so penetrating that it taints and renders useless everything it touches. Provisions even near it become uneatable, although the skunk's own flesh is white, and sweet, and even said to be delicious eating. Clothes saturated with it will retain the smell for several weeks, even though they are repeatedly washed and dried. A drop of the liquid in the eyes will cause blindness, and Indians are said not infrequently to lose their sight from this cause. Owing to such a remarkable power of offence, the skunk is rarely attacked by other animals, and its black and white fur and the bushy white tail, carried erect when disturbed, form what Mr. Wallace regards as danger signals by which it is easily distinguished in twilight or moonlight from other animals. Its sense that it needs but to be seen to be avoided gives it, he thinks, that slowness of motion and fearlessness of aspect for which it is remarkable.

We have already seen how, among the civet tribe, there is a form, the cynogale, specially modified for aquatic life. Among the weasels there is also a form thus modified, namely, the well-known otters. They are all long-bodied, long-tailed animals, with flattened heads and short limbs, which have webbed toes, furnished with small, blunt claws. Otters, of which there are some sixteen species, are spread over the whole earth save in the Australian region. Most expert swimmers and divers, feeding almost exclusively on fish, which, when captured, they bring to shore to devour, they are rarely met with far from water. The most extremely modified form is found nowhere but on the rocky shores of certain parts of the North Pacific Ocean, especially the Aleutian Isles, Alaska, and down to Oregon. It is clothed in beautiful soft fur, which is so valued that much danger exists of the absolute extermination of the whole species. They feed on clams, mussels, and sea urchins, of which they are very fond, and which they break by striking the shells together while held in each fore paw, sucking out the contents as they are fractured by these efforts; they also undoubtedly eat crabs and the juicy, tender fronds of seaweed, and sometimes, no doubt, also fish.

The only predaceous animals which it now remains to refer to are the creatures of the dog tribe—the dogs, jackals, wolves, and foxes. The structure of these animals is so well known, and they are all formed so almost completely on one type, that but few words need here be said about them. They are all strictly digitigrade animals, with long muzzles and two tubercular teeth behind each sectorial tooth, except the long-eared Cape dog, which has more, the dholes, which have one less below, and the bush dog, which has one less on each side, both above and below. All have five toes to each fore paw and four toes to each hind paw, except the hyena-dog, which has no more than four toes to any of its feet. There are above thirty-five species of the dog tribe, and some are found naturally in every quarter of the globe, except that the dingo of Australia may have been artificially introduced there. The fox is cosmopolitan, except as regards Australia and South America. In the Old World, species of the dog tribe are found from Spitzbergen and Siberia to the Cape and Java; and in the New World, from the shores of the Arctic Ocean to Tierra del Fuego and the Falkland Islands. They are, however, not naturally natives of Madagascar, the West Indies, New Zealand, Celebes, the Philippine Islands, or Ceylon. The wolf, fox and Arctic fox are common to both the Old World and the New. Of the remaining thirty-two species, twenty belong to the former, while only twelve are peculiar to the latter.

Such, then, are the groups into which the existing beasts of prey may be divided, and thus we may come to apprehend "what is a racoon," as far as science yet enables us to answer that question. We find that carnivorous animals are made up of a dog tribe, a weasel tribe, a hyena tribe, a civet tribe, a cat tribe, a bear tribe, and a racoon tribe. Among all these the racoon and its allies holds a peculiar, because a more or less neutral, position. The dogs, the cats, and the bears present us respectively with three very marked distinct forms, and it has been proposed to divide the whole of the carnivora into these three groups, associating the civets and hyenas with the cats on the one hand, and the weasels and racoon group with the bears on the other. But when we turn to study those lithographs of past history—fossil remains—we find apparently connecting links between the dogs and the bears on the one hand, and between the dogs and the civets on the other; while yet others seem to connect the civets and the weasels. In this puzzling maze it seems we must provisionally rest contented with the groups indicated by the study of existing species, which are the only ones of which we can know more than the bones and teeth. Of these the dogs, cats, bears, and weasels seem each to constitute a very distinct group, nor do the racoons really resemble the bears, except in the number of their blunt or tubercular grinders, while the civets and hyenas do approach the cats. Thus we find the racoon to be a type of a very small independent group of beasts of prey, standing, as it were, in the midst of the dogs, cats, weasels, and bears, without showing a decided and unmistakable affinity with any one of them. As to its special relationship with extinct forms of life, as yet we know too little to venture upon any affirmation.

# IX

## THE SLOTH

"THE inertia of this animal is not so much due to laziness as to wretchedness, it is the consequence of its faulty structure. With thighs ill-articulated its legs are too short, badly formed and worse terminated. For it has no sole on which to rest its foot, nor thumbs, nor separately movable fingers, but only two or three very long claws, curved downwards, which cannot be moved separately, and are more inconvenient for walking than serviceable for climbing. Inactivity, stupidity, and even habitual suffering result from its strange and ill-constructed conformation. Having no weapons for attack or defence, no mode of refuge even by burrowing, its only safety is in flight. Confined within the narrowest range, only climbing with difficulty or dragging itself along painfully, never allowing its plaintive voice to be heard except at night, everything about it shows its wretchedness and proclaims it to be one of those defective monsters, those imperfect sketches, which Nature has sometimes formed, and which, having scarcely the faculty of existence, could only continue for a short time and have since been removed from the catalogue of living beings. Truly, if the sloths did not inhabit deserts, if man or powerful animals had multiplied where they have their abode, they would not have lived down to our days, but would have already met with that extermination which will befall them later. They are the last possible term amongst creatures of flesh and blood, and any further defect would have made their existence impossible. To regard these imperfect

THE SLOTH 247

sketches of animal life, as being as good as others; to admit final causes for such ill-proportioned creatures, and to find that Nature is as admirable in them as in her finest works, is to take a most narrow view of the world and make our own ideas of finality the tests of Nature's aims."

In this quotation we have a memorable example of the errors into which the greatest thinkers may sometimes fall. It records a rash judgment (with respect to the sloth) which the illustrious zoologist Buffon allowed himself to make, and which he has recorded in the thirteenth volume of his immortal "Natural History." When we recall to mind how sagacious a thinker the great French naturalist was, the luminous suggestions, far in advance of his time, which he often threw out—as for example in his general comparison between the animals of the Old World and the New—we may well wonder at his having written such a passage as that above cited. However as Homer, in the realm of poetry sometimes nods, so there is hardly a man of science or an historian who does not occasionally offer us some prosaic error. Thus Isaac Newton strangely boasted that he made no hypothesis, Linnæus classed together the walrus and the sloth, Cuvier fancied that from a fossil "foot" he could construct an extinct Zoological "Hercules." His restorations were indeed wonderful, but the principle he enunciated is none the less untenable. Moreover, he strangely failed to understand the true affinities of the barnacle, nor were pouched-beasts by any means correctly appreciated by him in spite of his zoological and anatomical genius. Our own "Prince of Anatomists," Owen, suffered ruefully from his failure to appreciate an ape's "Hippocampus Minor," while his vigorous opponent Huxley stood sponsor for that never-to-be-forgotten creature of the fancy " Bathybius."

Similarly, Buffon was led by his imagination, to be at once unjust to Nature and to such a marvellous produce of Nature as the sloth. That animal is truly no less admirable in its organisation than is any which Buffon would have regarded as being one of Nature's "finest works," and far from being an imperfect and tentative sketch of animal life, it is a fully completed study of perfect adaptation of structure to need.

The fact is, no animal can be correctly appreciated by us if we do not well understand the circumstances of its being, its surrounding conditions. Each creature's structure is an expression and manifestation of that interplay of influences and activities between its own being and its environment, which constitutes its life. Buffon mistook the sloth's organisation because he was ignorant of the nature of the region it inhabits, namely the vast forest region of South America. The creatures of that region are formed exceptionally for arboreal life, as we have already seen with respect to the spider-monkeys and howling-monkeys. In that immense continent of foliage, Brazil, we have indeed a land which has produced, as it were, a great symphony of organic harmony composed in the forest " key." There we find, specially modified for an existence amidst trees, many orders of animals which elsewhere are not so modified, and above all the sloth, which is a creature fitted for the forest, as the camel for the desert, the dolphin for the water, or the eagle for the air. The colour of the abundant verdure even gains upon the animal world itself, and snakes and lizards, frogs and insects wear a livery of green. Not only colour but even form, may be thus affected, and the strange leaf-insects crawl above each,

## THE SLOTH

being in limb and body apparently a perfect foliar fragment.

The sloth then is an animal specially formed to dwell nowhere but in luxuriant forests and to feed exclusively on the leaves of trees, young shoots and fruits. Such food is there abundantly and perennially supplied on every side. Therefore it has not the slightest need for rapid motion to obtain it, and it would evidently be an economy, were it enabled to remain permanently in the midst of such abundance without the necessity of descending to the ground. And this is just what the sloth is enabled to do. He dwells, as it were, in a palace of many chambers, lined with beautiful hangings and many ornaments, all of which are good to eat, and all of which, after being eaten, are replaced as if by magic, to serve later for another repast, and so on without limit.

But to live thus, ever high up amidst the leafy branches of the forest and to dwell there securely by night and day, while being at the same time devoid of the activity of monkeys and other such arboreal beasts, necessitates a special and peculiar structure. Evidently less call is made upon the vital powers of an animal if it hangs passively, than if it has to hold itself up actively upon the trunk or branches of trees. The whole organisation of the sloth is dominated and governed by this need, the need of hanging passively and permanently, without any exertion or effort, from the branches of the trees amidst which it lives. But evidently, again, it is impossible that an animal formed to do this, can at the same time be organised so as to move well and freely on the surface of the ground, for which the stress and leverage must be altogether different. Hence the structure of such a creature must seem very defective to any one who only observes its motions on the surface of the soil, a position in which it naturally hardly if ever finds itself. Hence arises the apparent disproportionate length of its arms compared with that of its legs, and also the seemingly most defective conformation of its hands and feet. Sloths pass their lives hanging under the branches of trees, back downwards and so they can sleep securely. There they continue to hang even after death, till decomposition has far advanced, as any collecting naturalist may chance to find to his great disappointment. Sloths are difficult of detection, partly on account of the slowness of their movements, but more on account of their external appearance; for they are clothed with dry shaggy hair, often of a greenish tint, so that they are by no means unlike the masses of moss and lichen with which the forest trees abound. This green tint is not due to the colour of the hair itself, but to a minute algoid plant which lives upon the hair of the animal, the surface of each hair being peculiarly grooved. The growth of this small plant is also further favoured by the excessive dampness of the gloomy tropical forests, and it soon disappears from the hair of animals kept in captivity in England.

There are certain small points of structure wherein the sloth more resembles a reptile than a beast, and it is also reptilian in its great tenacity of life and the persistence with which muscular movement can be induced in the body after death. It will survive the most severe injuries, and can be given large doses of poison with impunity. Thus the sluggishness of its nature seems to extend into the very substance of its body. Therewith it is naturally a most inoffensive animal, nocturnal in its habits, solitary and almost always silent. The female has usually but a single young one at a time.

Many an arboreal animal is furnished with a prehensile tail, but the tail of the sloth is quite rudimentary. Being thus deprived of one mode of prehension, it is necessary that its other means of clinging should be all the more trustworthy. And nothing could be more trustworthy than those of the sloth, which consist only of its hands and feet, each one of which is so modified as to be, practically, but a somewhat movable hook. This is due to the fact that the fingers and toes of each hand and foot are so closely bound together that they cannot be separated; while each finger and toe is furnished with an enormously long and very strong nail greatly curved. When at rest, the hands and feet are so bent that each thus forms a strong hook, and it requires an effort on the part of the animal to unhook either a hand or foot from the branch it clasps. Thus it is that the sloth can sleep suspended from a branch, and remain so after death. But the fore paws can grasp and carry to the mouth, fruits, twigs with leaves or other objects, so that these paws do answer the purpose of hands in spite of their fingers being so closely bound together.

We have hitherto spoken of "the sloth," as if there was but one kind. There are however several kinds which form two distinct groups or "genera." The first genus contains the species known as the unau, or two-toed sloth, because it has only two fingers, fully developed and with long claws, to each hand; these two fingers answer to the index and middle fingers of the human hand. In captivity, the unau will eat bread and milk, vegetables and fruits, either cooked or raw. Its voice has been compared to the bleat of a sheep, but is seldom heard. It will also snort violently when seized.

The sloths are very exceptional with respect to the

bones of the neck. Every other mammal besides them —with two solitary exceptions—has the same number of these bones. Whether the neck is extremely long, as in the giraffe, or extremely short as in the mole, and still more so relatively in the whale, and porpoises, seven

FIG. 68.

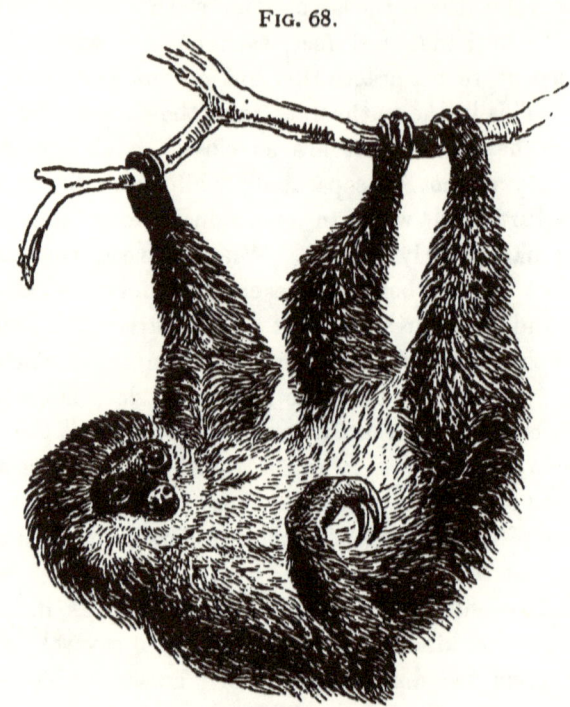

THE TWO-TOED SLOTH.

neck-bones and no more are always present. The same is also the case with man. One variety of two-toed sloths has also seven neck-bones, but a closely allied form, called Hoffman's sloth, has only six. This latter is the more northern kind of unau, and it extends from Ecuador, through Panama, to Costa Rica.

The other genus of sloths contains the "ai," or

three-toed sloth, the hand of which is a triple-hook, formed of three digits (each armed with a long strong claw), which answer to the three middle fingers of the human hand. There are several varieties of the three-toed sloth, but they all agree in one more exceptional character, since no other beast whatever possesses it. This is the possession of no less than nine bones, or "vertebræ," in the neck. What may be the reason of this strange peculiarity we are quite unable to conjecture. The ai is generally a silent animal, but an individual in captivity is recorded to have emitted, when pulled away from a branch to which it was clinging, a shrill note like that of some monkeys. The teeth of the sloths are exceedingly simple in form, and there are none whatever in the front of the mouth. There are usually but five above and four below on each side, and only in the unau is each first tooth prolonged beyond the others. The stomach of these leaf-eating animals is complex, reminding us of that of ruminating animals.

The windpipe of the ai does not pass straight downwards and backwards to the lungs as it does in all other mammals. It is folded on itself in a coil as is often the case in birds and reptiles.

Such are the main points which may here be mentioned, with respect to the structure, appearance, and habit of these South American beasts; but what, after all, are sloths? They have sometimes been supposed to be distant relations of the monkeys. They were for a time thought to be so by the great Linnæus, and the modern distinguished French naturalist, De Blainville, also so considered them. With their round heads and long arms they may be said to possess a certain resemblance to some of the apes, but such resemblance is indeed of the most superficial kind. Most naturalists have associated them with two other groups of animals which are also exclusively American, namely, the ant-eaters and armadillos, and these again with certain Old World ant-eaters, namely, with those known as pangolins (or the manis), and that called the Cape ant-eater, earth pig or aard-vark of the Dutch Boers. The whole of these form one order of beasts called "Edentata," because they have either no teeth, or at least none in the front of the mouth, while their teeth, even

FIG. 69.

THE GREAT ANT-EATER.

if numerous, are of a peculiarly simple structure, save in the aard-vark, in which they are complex, but complex in a way found in no other kind of beast whatever.

Thus the edentate order of beasts may be taken as a group parallel to the other orders of mammals, such as those of the apes, bats, carnivores, hoofed-beasts, whales, and porpoises, rodents, and insect-eating beasts, respectively.

Of the American ant-eaters there are three very distinct kinds, and they are singularly different in appearance and habits from the sloths. The great ant-eater stands two feet high at the shoulders, has a very long and bushy tail, and measures four feet from its root to the end of the snout.

Nothing could well be more unlike the head of a sloth, than is the head of the ant-eater, which is drawn out into an exceedingly long and slender snout with a small mouth, which opens at its extremity only so far as to allow a very long worm-like tongue to be protruded from it. The animal is to be found far and wide, though it is nowhere plentiful, in the tropical parts of South and Central America, in damp forests and the vicinity of rivers and swamps. It has claws on its fingers save the fifth, but that of the middle finger is extremely large and strong. It uses its claws to open the nests of white ants, or termites. Then as the insects rush out when their nest is broken into, the ant-eater rapidly introduces its long tongue amongst them, to which they adhere, because it is coated with a very glutinous saliva, secreted by enormous spittle-glands. The tongue is sent forth and drawn back with great celerity, and thus the animal is enabled to obtain a great quantity of the small creatures, which constitute its main food in a state of nature. It is entirely toothless, but has a strong muscular stomach, like the gizzard of a bird. In captivity, it will eat bread and milk, also blood and newly-born rats. It does not climb nor does it burrow, its claws being only used for tearing down ant's nests, and for defence, in which latter action it can use them very effectively. When not attacked, it is, however, an inoffensive animal. It has but a single young one at a birth.

The middle-sized ant-eater, or tamandua, has the characters of the larger one less developed, but its tail is not at all bushy, and it can climb trees. The third and smallest ant-eater is entirely arboreal, of a yellow colour and about the size of a rat. It is called the two-toed ant-eater, because it is only the second and middle fingers which have claws. Like the tamandua it is only found in the forests of Central and South America.

FIG. 70.

THE SMALLEST ANT-EATER.

The ant-eaters are different indeed in aspect from the sloths, but the next group, the armadillos, are still more divergent.

They are amongst beasts what tortoises are amongst reptiles; inasmuch as a large part of their skin is ossified, that is, changed into a true bony substance which is everywhere externally invested by horny scales. Almost always this external structure consists of a solid shield on the head, one over the shoulders, and one over the hinder portion of the body; the back and sides —between the shoulders and hinder plates—being invested by transverse solid bands (the number of which varies with the species), which are connected with each other by soft, flexible skin, so as to allow the body to be bent, and indeed, sometimes to be rolled up into a ball, the soft ventral surface of the body being by this means concealed, and the hard solid coat presented against attack on every side.

Armadillos are small or moderate-sized animals, which are mostly nocturnal, and feed on both animal and vegetable substances; eating insects, worms, reptiles, roots, and carrion. They are powerful and rapid burrowers, by which faculty alone they can escape their enemies, for they are not only harmless but defenceless—save as regards their armour—and offer no resistance when caught. As to them, Darwin tells us in his journal during the voyage of the *Beagle* :

"In the course of a day's ride near Bahia Blanca, several armadillos were generally met with. The instant one was perceived, it was necessary, in order to catch it, almost to tumble off one's horse : for in soft soil the animal burrowed so quickly, that its hind quarters would almost disappear before one could alight. It seems almost a pity to kill such nice little animals, for, as a Gaucho said, while sharpening his knife on the back of one, 'Son tan mansos' (they are so quiet)."

Most of the species, however, are prized as food.

In order to burrow, it is necessary that they should move their fore limbs rapidly, and with much force. It is therefore necessary that the muscles of the chest should be largely developed, and in order that there should be more space for such muscles, the breast develops a keel, as it also does in the mole and in bats, and to a very much greater extent, in birds—as every one who has carved a fowl must know.

The largest of the armadillos—the priodon—measures a yard from snout to tail root, and the tail itself is twenty inches long. It is found in the forests of Surinam and Brazil, and has very powerful claws. It is accused of burrowing in graveyards and feeding on dead bodies, but

FIG. 71.

THE APAR ARMADDILLO.

it lives principally on white ants and other insects. It has no less than from twelve to thirteen movable bands interposed between its anterior and posterior body-shields. Another kind known as the apar has but three movable bands, but it can roll itself up very perfectly. Darwin tells us:

"It has the power of rolling itself up into a perfect sphere, like one kind of English woodlouse. In this state it is safe from the attack of dogs, for the dog not being able to take the whole in its mouth, tries to bite one side, and the ball slips away. The smooth, hard covering offers a better defence than the sharp spines of the hedgehog."

Armadillos dwell in the open plains and also in the forests of South America, but one kind, the peba or nine-banded armadillo, ranges from Paraguay to Texas. The commonest kind is the six-banded armadillo, or encoubert, which inhabits Brazil and Paraguay, and allied forms, more or less hairy, are found south of the Rio de la Plata, and another still more hairy species is found in Peru.

A small, very rare, and peculiar kind, is the pichiciago, of which one variety is found in the Argentine Republic, and another in Bolivia. Its bony plates are very thin and delicate, and form no solid shield over either the shoulders or the haunches. The horny covering of the Argentine species is of a pinkish colour, and the animal has snow-white, silky hair.

All the armadillos have simple teeth, and the number on each side of either jaw may vary from seven to five-and-twenty.

Very different from the armadillos of America, are the pangolins of the Old World, although they also are protected by a dense and strong external armature. Their skin, however, contains no bony plates; their body being covered and protected, except on its under surface, by large, close-set, horny, over-lapping scales, amongst which grow a few hairs. Indeed, the scales themselves—strong and dense as they are, with sharp cutting edges—are really composed of hairs cemented together. The limbs are short, but the tail is moderately, or greatly, elongated according to the species, of which there are some seven kinds. One African kind, appropriately called the long-tailed manis, has a greater number of bones in its tail than has any other beast—namely forty-nine. Pangolins can roll themselves up into a ball, when the sharp edges of their scales standing more or less out on all sides, effectually protect them from attack. They feed, as do the ant-eaters of the New World; and their jaws are equally toothless, while their fore-paws are furnished with powerful claws, especially that of the middle finger. They both climb and burrow, sometimes forming a chamber six feet wide, four yards beneath the surface of the soil; the size of course depends on the size of the species, and they vary from one foot to five feet in length. They are said to live in pairs, and have one or two young at a birth. The only sound they give forth is a hiss, and

FIG. 72.

THE LONG-TAILED PANGOLIN.

this, together with their scaly exterior, naturally led to their being considered to be reptiles—some kind of lizard. They are found in China, India, and the Malay Archipelago, and Africa south of the Sahara. No species is now common to both Asia and Africa, though in quite recent geological times, a form now African, existed at Madras. At an earlier date a species dwelt in South-eastern Europe, which was three times the size of any species now living.

Mr. Fraser, the African travelling zoölogist, gives * us some details as to the habits of an animal of this kind. He says:

" During my short residence at Fernando Po, I suc-

* In his "Zoologica Typica."

## THE SLOTH

ceeded in procuring two living specimens of this animal. The individuals were evidently not adult: the largest measuring thirty inches in length, of which the head and body were twelve inches and the tail eighteen inches. I kept them alive for about a week at Fernando Po, and allowed them the range of a room, where they fed upon a small black ant, which is very abundant and troublesome in the houses and elsewhere. Even when first procured they displayed little or no fear, but continued to climb about the room without noticing my occasional entrance. They would climb up the somewhat roughly hewn square posts which supported the building with great facility, and upon reaching the ceiling would return head foremost; sometimes they would roll themselves up into a ball and throw themselves down, and apparently without experiencing any inconvenience from the fall, which was in a measure broken upon reaching the ground by the semi-yielding scales, which were thrown into an erect position by the curve of the body of the animal. In climbing, the tail, with its strongly pointed scales beneath, was used to assist the feet, and the grasp of the hind feet, assisted by the tail, was so powerful, that the animal would throw the body back (when on the post) into a horizontal position, and sway itself to and fro, apparently taking pleasure in this kind of exercise. It always slept with the body rolled up, and when in this position in a corner of the building, owing to the position and strength of the scales, and the power of the limbs combined, I found it impossible to remove the animal against its will, the points of the scales being inserted into every little notch and hollow of the surrounding objects. The eyes are very dark and hazel and very prominent. The colonial name for this species of manis is 'Attadillo,' and it is called by the natives of the island 'Gahlah.' The flesh is said to be exceedingly good eating, and it is in great request among the natives."

The American ant-eater we found to be very unlike sloths, and the armadillos are still more unlike them. The pangolins would seem fully as unlike them as the armadillos, and, nevertheless, there is one very strange character present in some pangolins, which forcibly brings the sloths to mind. We have already called attention to the very remarkable fact that sloths differ from all other beasts save two, in that they have neck-bones which may be as few as six, or as many as nine instead of being of the sabbatical number, seven, as they are in mammals, with both the longest and the shortest necks. Now, it has been of late years discovered that in some pangolins there are eight bones in the neck. If there is no really exceptional affinity between the pangolins and sloths, the fact, that they thus agree to differ from all other beasts (save one) and man, in the number of their neck-bones, is a very interesting one.

That it may be a mere coincidence is, however, rendered less improbable by the fact, that the only other creature which exhibits a divergence from the normal condition of beasts, as to this character, is the manatee, an aquatic animal which certainly has no special relationship to either sloths or pangolins.

The last group of that order of animals in which the sloth is classed consists but of two species of earth-pig, or aard-vark, one of which inhabits South Africa, while the other is to be found in North-east Africa, including Egypt. In tertiary times, a species also existed in what is now the Island of Samos in the Turkish Archipelago These creatures differ greatly from other edentates, while in the structure of their teeth they diverge in the most remarkable manner from every known mammal, and approximate to certain fishes.

The earth-pig has at first sight a rough resemblance to a large hog. It is scantily covered with bristle-like hairs, has a long snout, the end of which is very mobile, and with terminal nostrils, long ears, rather short limbs, and a long tail. It measures about three and a half feet from the root of the tail to the end of the snout; the tail is a foot and three-quarters long, and is very thick at the base, but tapering towards the point. There are four toes to each fore-foot and five to each hind-foot, all provided with large strong claws, flattened horizontally and hollowed out underneath. The tongue is thick and fleshy, and much less vermiform than that of the ant-eater and pangolins; but the spittle-glands are still,

FIG. 73.

THE AARD VARK.

as in them, largely developed. It used to be an extremely common animal in South Africa, though from its nocturnal habits and timidity it was never frequently seen. It feeds on white ants almost exclusively, so that when these insects are very abundant, the presence of an aard-vark may be anticipated.

The termites raise mounds of an elliptical figure to the height of three or four feet above the surface of the ground; and so numerous are these gigantic ant-hills in some part of South Africa, that they may be seen to extend over a plain as far as the eye can reach, and so close together that a waggon may have a difficulty in passing between them. From the action of the sun, these nests become exceedingly hard on the surface. It is in the neighbourhood of such nests that the aard-vark makes a deep burrow, in which it sleeps during the day, and it burrows with such extreme ease and celerity, that it is said to be a hopeless task to try and dig the animal out, while it is so strong that if caught it takes two or three men to drag it out of its burrow. At night it goes out to one of the nearest ants' nests, and scratches a hole in the side of it big enough to admit its snout, and then rests, quietly inserting its tongue again and again into the aperture it has made, withdrawing it each time covered with ants which have flown out to defend their dwelling, and have been caught by the sticky saliva with which its tongue is coated. It becomes very fat, and its flesh is esteemed both wholesome and palatable, the hind quarters, cut into hams and dried, being especially relished. It may seem strange that so bulky an animal should get fat on such food. But termites are practically infinite in number in the tropics, and may attain a length of from one inch to one inch and a half. Their bodies also are soft and unctuous, and are often collected and eaten by the natives of Africa. The traveller, Paterson, affirmed that only prejudice prevented Europeans from making a similar use of them, and declared that in his different journeys, he was often under the necessity of eating them, and that he found them far from disagreeable, while farmers collect them by bushels, for the purpose of feeding poultry.

It is difficult to detect any relation between such food and even the external form of the teeth of the aard-vark, and absolutely impossible so to explain their structure. One does not see why the creature should require teeth any more than the pangolins and ant-eaters which also feed on termites.

There are usually five teeth on either side of each jaw, simple in shape, with crowns, which at first are rounded, but soon wear flat. Careful inspection of the worn surface shows a number of small holes which are the apertures of as many canals, and the tooth seems to have the structure of a cane. In fact, however, each tooth, though apparently simple, is really composed of a closely set bundle of very fine, long, cylindrical teeth, united together side by side. As the reader no doubt knows, each of our own teeth has a soft centre, known as the "pulp," by the hardening of the outer part of which each tooth has been formed. The very fine canals, which run through the substance of each tooth of the aard-vark, are the cavities for such pulp of the various very long and slender teeth, by the fusion of which each apparently single tooth is formed. Such a structure exists in no other beast, and even in no other reptile; but, strange to say, the same thing is found in a fish of the skate kind, known in science as "*Myliobatis.*" Yet the aard-vark can have no special relationship of generic affinity with these fishes.

It would be impossible to find a stronger instance of that circumstance on which, in our previous articles, we have so often insisted, namely, the independent origin of similar structures.

Such are the beasts that live on the earth's surface to-day, and compose the singular, and singularly diversified, order of Edentates.

Outside that order, we have already met* with two animals which have been supposed to be therewith

* See pp. 49 and 50.

allied; though the resemblance is but an external one, and, so far as it goes, yet further illustrates the doctrine of the independent origin of similar structures. These two animals are the echidna and the native sloth. The echidna has the powerful claws, the long snout, the vermiform tongue, and the copious spittle-glands, fitted for a creature which is by habit an ant-eater. Nevertheless, in essential structure, it is poles asunder from the true ant-eaters; for as we saw in our article "The Opossum," the echidna and platypus together constitute the most aberrant of all the groups which compose the class of beasts, namely, the group known as "Ornithodelphous mammals," which form a sub-class by themselves.

Similarly, the native sloth, or koala, is a creature which resembles the true sloth, because it also is formed to pass the greater part of its life clinging to the branches of trees. It lives on the tender shoots of trees and on leaves, and has a round head, long claws and short tail, and it is also very tenacious of life, and even when severely wounded will not quit its hold of the branch to which it may be clinging. But however many superficial points of resemblance the koala may have to a sloth, it is a most different kind of animal, for as we have seen, it belongs to the sub-order of pouched-beasts, or Marsupials; while the sloths are what we ourselves are, namely, Monodelphous mammals. Therefore, whatever structural resemblance may exist either between the echidna and the ant-eater, or between the koala and the sloth, must be resemblances which have been induced—arisen independently—and have never been inherited from a common ancestor.

But if the sloth has no affinity to any animals outside the Edentate order, has it any special affinity to any of the other animals within it? Of the ant-eaters, armadillos, pangolins, and aard-varks, which group is the least divergent from the sloth—an animal apparently so different from them all?

As we have seen, that most exceptional character, a divergence from the number seven in the bones of the neck, would seem to connect the pangolins with the sloths. But the resemblance is an absolutely isolated one, and is accompanied by a great number of differences, not the least important of which is the geographical difference, since they respectively inhabit the old and the new worlds only.

The aard-vark in various points, but above all in tooth structure, is so divergent from all other Edentates, that it is quite impossible to recognise in it a creature with any special relationship to the sloth. It again also differs from the latter, in being an inhabitant of the east side of the Atlantic.

Armadillos, like sloths, are exclusively American creatures, and they have simple teeth, whereas the ant-eaters differ from the sloths in having none. But, with this exception, it is impossible to detect any special resemblance between these beasts, essentially terrestrial, and burrowing and clothed in bony armour, and the arboreal sloths, clothed with their coarse hair.

There only remain, then, the ant-eaters wherewith to compare the sloths, and they are different from them indeed. As before said, nothing could well be more different from the round-headed and tooth-provided sloths, than the very long-snouted edentulous ant-eaters. The existing ant-eaters, then, afford us no clue whatever to any relationship which may exist between the sloth and any animal which is not a sloth. We must therefore turn to those organic records of the past, which remain, in order to see if there we can find evidences such as do not exist in the living world, to guide us in our quest.

In the first place, it may be mentioned that, in the latest tertiary cavern-deposits of Brazil, and in the vicinity of Buenos Ayres, remains have been found of an enormous kind of armadillo, which differed greatly from all existing armadillos. Its bony coat was entire, reaching from the neck to the root of the tail, without the interposition of even a single movable band; the tail was also enclosed in a strong bony sheath. Many varieties of these animals, known as Glyptodons, once existed, and some have left their remains not only in Mexico, but also in Texas. But though these animals diverge in structure from existing armadillos, they do not diverge from them in the direction of the sloth, but rather in carrying to a still higher degree the characters of the armadillo type. This makes them especially interesting, because, as a rule, the animals which are most recent are most specialised, while earlier forms are more generalised in their organisation. Such was by no means, however, invariably the case, and we have a remarkable instance to the contrary, in the sabre-toothed tigers, which, as is shown in our article on the racoon—carried the carnivorous organisation to a higher degree of specialisation and perfection than do any tigers, lions, or other members of the cat-tribe which exist in our own day. Not only, indeed, did these ancient armadillos have a shell or carapace in one solid piece, but the backbone, instead of consisting of a series of distinct bones (vertebræ) separated by joints, had their vertebræ almost entirely fixed together into one solid bony tube, containing the spinal marrow, with one complex joint at the base of the neck to allow of the head being withdrawn within the shelter of the carapace.

It is a singular thing that an animal so strangely built, and which could, one would think, defy the attacks of almost any and every enemy, should have become extinct.

But if the glyptodons fail to offer us any indication of the line of descent along which the sloths have travelled, such is not the case with another most singular and important group of animals, which have scattered their huge remains broadcast over both North and South America.

In that, politically, most eventful year, 1789, there arrived at the Royal Museum, in Madrid, an almost perfect skeleton of a huge beast, the bones of which had been found on the banks of the river Luxan, near Buenos Ayres. The creature was of vast bulk, exceeding in size every existing land-animal except the elephant, and had its limbs been larger it would have surpassed even the dimensions of the elephant. It measured full thirteen feet from the front of the head to the end of the tail, and it had a very strong tail, which was itself five feet long. It inhabited those lands, the waters of which run into the Rio de la Plata; and similar forms have been found in South Carolina and Georgia. The animal is known as the Megatherium, and strange to say, in spite of its immense bulk, and in spite of its having been organised for walking on the ground, it was so like the subject of this paper, that it was called a ground-sloth. Such is especially the case with respect to certain points in the formation of the skull, the shoulder-blade, and the haunch bones. The fore-limbs were larger than the hind-limbs, and had three immense claws attached to its three middle digits, while there appears to have been but a single large claw in the hind foot, which claw was attached to the third toe. It seems to have walked on the outer sides of its hands and feet, and so the claws were not in contact with the ground, and thus could be kept sharp. Its teeth were essentially like those of the sloths, but rather more complicated, so that it probably ground up and devoured the smaller branches of trees, as well as their leaves and tender shoots. Another somewhat smaller South American beast, now extinct, is known as the Mylodon, of which there were a number of large varieties, while the remains of allied creatures have also been discovered in America, and are known as the Scelidotherium Megalonyx and the Nothrotherium, and these were all intermediate in structure between the existing sloths and ant-eaters, having the head and teeth of the former, with the trunk, and, in some respects, the limbs of the latter. Much speculation has taken place, as to how such huge creatures as the megatherium and mylodon could have succeeded in browsing on the leaves of trees, which, one would think, must have been completely out of their reach. It was at one time thought by some persons that they, like the sloths, actually climbed trees, and lived in the branches, and the suggestion was made, that, in their day, trees gigantic enough to have been in proportion to their size, might have existed. But there is no need for so wild an hypothesis, in favour of which no fragment of evidence exists. Doubtless their bulk enabled them to reach the lower branches of many trees, when the fore part of the body was raised and supported on the massive hind quarters. Then their great hook-like claws, at the end of their rather long and more or less prehensile fore-limbs, were no doubt very efficient organs for tearing down branches and cutting or breaking off smaller portions. But Sir Richard Owen has suggested yet another mode by which they might have obtained their leafy food still more abundantly. The development of the bones of their limbs makes it evident that those limbs were clothed with prodigiously voluminous, and therefore very powerful, muscles. This fact is shown by the huge ridges, and other prominences which stand out from the bones, and served to afford additional surface for the attachment of such vast muscles. The great haunch bones tell the same story, and the tail itself must have been a most powerful organ, if not for active movement, yet for most efficient support. Activity was doubtless no attribute of these huge beasts, which had nothing to fear from any enemy. Probably their movements were little, if at all, more rapid than are the movements of the sloth in our own day. But speed was not required for the mode of procuring food, which the venerable naturalist Owen has suggested. According to him, these creatures raised themselves nearly erect, supporting their ponderous body upon their two bulky hind limbs, and their very powerful tail, as on a tripod. Then with their strong fore-arms, they embraced the trunk of some moderate-sized tree, and proceeded to sway it to and fro, till they succeeded in prostrating it. Thus they would be provided with an ample repast, and their great cutting claws would be most useful for tearing down, or breaking off, those branches of the prostrate tree which were out of their reach, without such action on their part. The single great hooked claw attached to each hind foot was no doubt of great use to them, by giving them a much better and more secure hold on the ground, while struggling to uproot a tree, than they would have had without it.

We may now attempt to answer the question, "What is a sloth?" in the light afforded us by our brief review of those forms living and extinct, which naturalists have grouped in the same order with it—the order *Edentata*.

Of its five groups (1) sloths; (2) ant-eaters; (3) armadillos; (4) pangolins and (5) aard-varks, the fifth and last stands wide apart from the other four, and shows no sign of real relationship with any one of them. The aard-vark is placed with the other edentates rather because naturalists do not know in what other order to put it, and shrink from erecting it into an order by itself.

The pangolins also show little affinity to the first three groups. We have seen how even an animal, of so fundamentally distinct a nature as the echidna, may nevertheless present us with ant-eater-like characters, which cannot have been due to inheritance, but must have arisen independently. The ant-eater-like character of the pangolins, then, may also have arisen independently. Nevertheless, the very singular fact, that the bones of the neck have a tendency to diverge from the number almost universal in the class of mammals, seems to indicate a possibly deep-seated affinity between them and the sloths which present that abnormal character in a yet more marked degree. Nevertheless, we can only speak of it as a "possibly deep-seated affinity," because, as before said, we find similar abnormality in the manatee, which can hardly be supposed to have any exceptional affinity with either the sloth or the pangolins.

The armadillos do show some evident marks of affinity to the ant-eaters, especially in the structure of the lumbar region of their backbone. Nevertheless, in their external defensive armature, they differ widely from them, and from every other kind of beast whatsoever. Moreover, the most ancient form of the group yet known to us—the glyptodons—were even more exceptional in structure than those which exist to-day.

In that extensive group of huge creatures which, since it contains the megatherium and all creatures like it, may be spoken of as the group of "Megatherioids," we find animals which, at one and the same time, resemble both ant-eaters and sloths. On the principles of evolution then we may regard them as at least nearly allied to that parent form in which both sloths and ant-eaters had their first origin. From some more or less megatherium-like animals there was gradually evolved a creature on the way to elongate its snout and tongue, lose its teeth, augment its tail, and live on animal food. From such a form there must also have been gradually evolved some creature on the way to dwindle strangely in size, to shorten its tail, to simplify its teeth, to elongate its arms, and to diminish the number of its toes, which grew more and more rigid as the animal's race took more and more to climb up the trees, no single branch of which it could any longer hope to be able to pull down. Meantime the troops of great megatherioid animals fell off in numbers, and finally disappeared from the earth's surface (as the glyptodons have also done) and in an as yet quite inexplicable manner, leaving in their place the groups of smaller creatures, different indeed as to size, but more exceptionally and specially diversified in structure, than were the gigantic ancestors from which they may boast to have sprung.

A sloth then is the animal of all beasts the most exclusively organised to dwell in trees and live upon their foliage. It is one of the last evolved of a group of beasts, the most remote and unknown ancestors of which may have given origin to the pangolins and possibly also the aard-vark. Other unknown ancestors less remote were the source whence sprang the parents of the armadillos on the one hand, and the megatherioids on

s

the other: such megatheriods being the only yet known representatives of the parents of both the sloths and also of the ant-eaters, which we thus know to be the nearest allies of, although wonderfully different in habits from, that most exceptional and at first most misunderstood animal, the sloth.

## X

## THE SEA-LION

THE sea-lion is a beast the sight of which must be familiar to very many Americans, for it can hardly remain long unknown by any visitor to San Francisco. It is also a creature well worthy of notice in itself, since, interesting as sea-lions are from their habits, they are also remarkable as regards their structure.

A sea-lion is a marine mammal and a true quadruped, all the four limbs of which are modified to suit its eminently aquatic life. It has a moderately rounded head with large eyes, very small but quite distinct external ears, a longish neck, a long tapering body, and a very short tail. It has six small cutting teeth above, and four below in the front of the mouth, and external to them, on either side, is a large eye-tooth, or "canine," which is conical, pointed, and recurved. Behind these there are, on each side of the mouth, five or six grinders above and five below; each of which has a crown with three conical prominences, whereof the medium one is very much the largest. In the fore-limb the parts which answer to our upper arm and our fore-arm are exceedingly short, but the hand is very much elongated, with five fingers, which are all enclosed in one fold of skin, forming a fin, the separate elements of which are not visible externally. Indeed, the fold of skin which encloses the fingers extends much beyond their respective tips. Of the five fingers included within it, the little finger is by far the smallest, and thence they increase in length to the thumb, which lies in the same plane as the others, but is much the longest and the largest of them all. In the foot, the digit which answers to our great toe is also the longest and strongest one, but the other four are nearly equal in length to each other and to it—decreasing from the second to the fifth digit, very much less rapidly than do the corresponding digits of the hand. The five digits of the foot seem at the first glance to be much less completely involved in a common fold of skin than do those of the hand; but in fact they are all thus included to beyond their tips. The deceptive appearance of distinctness is due to the fold of integument which extends beyond them, being drawn out into five long processes of skin, one opposite the tip of each toe. The nails on the dorsal surface of each extremity mark the end of the respective digits. These nails are mere rounded rudiments, save in the three middle fingers of each hind-foot, each of which has an elongated, compressed, and curved claw.

The length of the animal from the nose to the root of the tail is often above six and a half feet; the tail is five and a half inches. The fore-limb measures about two and a half feet, and the hind-limb two and a quarter. Sometimes the animal may be a little over eight feet from the snout to the end of the tail.

The colour varies from chestnut-brown to blackish-brown, and there are long yellowish-white whiskers on either side of the muzzle.

The creature is a vivacious, active animal, even on land, and wonderfully graceful and quick in its movements in water, where its four limbs act simply as fins. On land it walks on all-fours, with the palmar surface of its hands and the plantar surface of its feet on the

## THE SEA-LION

ground. Then the digits of the hand are directed backwards, while those of the hind-foot are turned forwards. The two hind-limbs, however, are bound together by the skin almost down to the ankles, and so, in walking, that part of the body presents a singularly constrained and awkward appearance.

We have said that the sea-lion can hardly be unknown by any visitor to San Francisco, and such is indeed the case, for their destruction has been for some time

FIG. 74.

THE CALIFORNIAN SEA-LION.

forbidden, and a view of them (through a telescope) disporting themselves on some adjacent rocks, is one of the sights shown to those who come for the first time to the City of the Golden Gate.

During the last twenty years the animal has become well known even in Europe, and has been shown at the Central Park Menagerie, New York, and the Zoölogical Gardens of Philadelphia and Cincinnati. It has also been exhibited by the late Mr. P. T. Barnum. A sea-lion, though of a different variety, was first brought to the Zoölogical Society's Garden, in London, in 1866. It had been captured near Cape Horn in 1862, by a French sailor, who had taught it a great variety of tricks, which he made it play before admiring crowds in the London Gardens, where he had the care of it for several years. Two pairs of the Californian variety of sea-lions arrived in England in 1877, and various others have been received here and there on the Continent of Europe.

The animal which has thus become so familiar to us was, it seems, first made known by Captain William Dampier, who, in 1729, published his narrative of "A New Voyage Round the World," wherein is an account of observations on the sea-lion made by him in 1683.

The Californian sea-lion is distinguishable from other varieties by the marked angle which the forehead forms with the muzzle. It ranges the coast of California from San Diego and San Nicolas Island, to the Bay of St. Francisco.

The Northern or Steller's sea-lion, a creature about ten feet in length, inhabits the North Pacific. The Southern or Patagonian sea-lion comes from the Falkland Islands and Patagonia. A small kind of sea-lion is found at the Cape of Good Hope, while yet another variety, Forster's sea-lion, frequents the coast of Australia and various islands of the Southern Ocean, and has a delicate fur. Another kind, which yields the soft fur known by ladies as "sealskin," is often distinguished as the "sea-bear." It is an inhabitant of those tiny islands, St. George and St. Paul, which are two of the Prybiloff group of islands, whence its skins are imported in immense numbers.

These islands are toward the east end of Behring's Sea, but individuals are also to be met with in Behring's and Copper Islands, at the west end of Behring's Sea.

The Southern or Patagonian sea-lion was, in 1868, according to Captain Abbott, very common on the Falkland Islands, where it bred and was little disturbed by sealing boats

"There is a remarkable disparity," he tells us,* "between the male and female of this variety. The male is as large as a bullock in circumference, while the female is no bigger than a calf. At one time only the female was killed by the sealers, as the skin of the male was considered to be of little value; and this may account for the preponderance of males which I here observed. . . . . I recollect on one occasion, accompanied by a friend, rolling stones down from above on some that were lying on the beach. When one was hit, he gave a roar and rushed at his nearest companion, fancying no doubt that he had attacked him; others swallowed the stones thrown at them. . . . . Although these animals are so unwieldy in appearance, they have wonderful powers of climbing, chiefly by means of their flippers, and can ascend rocks that are almost perpendicular. I recollect once watching a number of seals from the top of a very steep ledge of rock about twenty feet high, when upon hearing our voices, a large sea-lion gave a sudden roar and rushed up the rock where I was sitting. I fancy that it was on account of a female companion near him that he made this attack, as among about fifteen males, there appeared to be only two females."

Forster's sea-lion seems to have been seen by Captain Cook off the north part of New Zealand, in January 1770, and later on he saw the largest animal of the kind he had ever beheld. It was swimming on the surface of the water and suffered its pursuers to come near enough to fire at it, but after an hour's chase it got clear away.

The sea-lions and sea-bears, both known by the common term of "Eared-seals," are almost wholly confined to

* "Proceedings of the Zoölogical Society, 1868," p. 191.

temperate and cold latitudes, the different varieties being found in the localities before indicated. Most of them are clothed only with coarse, hard, stiff hair, varying in length with age and the season of the year, and being wholly devoid of soft under-fur. But the sea-bear of the Prybiloff Islands and the southern sea-bear have, as before said, an abundant soft, silky under-fur, which gives to the skins, especially of the females and young males, great value as articles of commerce. In these precious skins, the longer, coarser over-hair also varies in length and abundance according to age and the season of the year.

There is generally a wonderful disparity in size between the sexes, similar to that noticed by Abbott in the Patagonian variety. Adult males may weigh three, and sometimes five times as much as do the females. They are all greedy devourers of fish, which their sharp-pointed, trident-shaped, grinding, or molar, teeth enable them to catch and retain easily. In confinement they will eat a prodigious quantity of fish, and readily learn to catch them in the air when they are thrown to them, as also to climb into and perch themselves on chairs, and to play a number of amusing tricks, the acquisition of which shows that they possess considerable intelligence as well as docility.

It is, however, very surprising that, in spite of their voracity, the males at least will, at certain seasons, remain for almost an incredible period without food, although at the very time they may be exerting great combative energy. This is connected with their breeding habits, which are very singular and remarkable. They are gregarious and polygamous animals, which associate together in vast troops.

Nevertheless, the troops are largely composed of

separate families, each under one powerful male who governs it. The places where these animals congregate are popularly known as "rookeries." When the breeding season approaches, the adult or old males come first to the rookery, and landing, each takes up his station on the rocks close to the sea, there to await the arrival of the females. But their respective stations are not taken up without severe contests, the most powerful males securing advantageous positions close to the sea, while those which have not reached their prime or which have survived it, are driven further inland. They fight desperately, and sometimes a male will have to carry on fifty or sixty successive contests before his position on the shore-line is fully secured to him. When the males arrive they are strong, vigorous, and exceedingly fat, and the successful ones secure a space for themselves about ten feet square. They begin to arrive in the latter part of May, and a little before the middle of June the first females approach the shore. This is the signal for a universal and desperate fight amongst the males. Each successful combatant then tries to coax or to force a female to land at his station, where she is immediately looked after with the most vigilant jealousy by her lord till he sees another female approaching, when he goes to the water's edge to similarly coax or force her. Then comes the opportunity for one of the males less favourably situated to appropriate to himself the female thus momentarily left unguarded. She is seized and deposited on his ground, but hardly is this accomplished before he has in turn to defend his conquest against the attempts of other males whose stations are still further back. Should any of them succeed, they in their turn have to combat for their prize, till the much-disputed fair one falls to the lot of a male sufficiently strong to maintain possession, and sufficiently remote not to yield to the temptation of neglecting to watch over her, in order to secure an additional bride. Meanwhile, the strong and vigorous males close to the shore will usually obtain from a dozen to fifteen wives, and as many as forty-five have been found appropriated by one powerful old male. The females never fight or quarrel with one another, and are said seldom to utter a cry of pain, although they are often severely wounded while being contended for by the males, who each seize them with their teeth.

The sea-lions fight almost entirely with the mouth, and are often covered with scars and gashes, not unfrequently losing an eye in their struggles. Mr. Elliott * tells us that they usually approach each other with averted heads and a great many false passes before either one or the other takes the initiative by gripping; their heads are darted out and back with the greatest rapidity, their hoarse roaring or shrill whistling never ceasing, while their fat bodies writhe and swell with exertion and rage, blood streaming from their bodies and their fur flying about in all directions. When one of the combatants feels he has had enough and retires, he is never pursued by his conqueror, who remains quiet, uttering a peculiar chuckle, which seems to indicate satisfaction and contempt, keeping, however, a sharp eye open all the time for the next rival who may approach.

Owing to their gregarious habits the females lie most contentedly together in the largest harems. The males during the breeding season remain wholly upon land, and they will suffer death rather than leave the spot they have chosen. They thus sustain, for a period of three

* See his report on the Prybiloff group of Fur Seal Islands of Alaska.

or even four months, an uninterrupted fast, being nourished wholly by the absorption of the fat of their own bodies, so that at the end of the breeding season —the beginning of August—they have become quite weak and emaciated. They have also to abstain entirely from water during this fast.

As to their courage and determination in refusing to leave a chosen station, Mr. Elliott repeatedly tried to drive them away, and to put their courage to a test, he one day walked up to within twenty feet of a male, which had four females with him, and peppered him with dust shot. His bearing, in spite of the noise, smell of powder, and the pain he must have felt, did not change in the least from the usual attitude of determined defence which nearly all the males assume when attacked with showers of stones and noise. He would dart out right and left and catch the females, which timidly attempted to run away after each report, and fling and drag them back to their places; then, stretching himself up to his full height, he looked Mr. Elliott directly and defiantly in the face, roaring and spitting most vehemently. He next made various little charges of ten or fifteen feet at his assailant, afterwards retreating to his old position which he would not go back from, seeming resolved to hold his own or die in the attempt. But though thus courageous and persevering in defence, the sea-lion never took the offensive beyond the boundary of his station, so far as Mr. Elliott observed.

One pup is born at a time, and the mother's milk is abundant, rich, and creamy. But she seems very apathetic with her offspring. The observer before referred to never saw a female caress or fondle her cub, and if it had strayed but a short distance beyond the bounds of the harem, it might be killed before the mother's eye without causing her to show the slightest concern. The same indifference is exhibited by the male to what takes place outside the space of ground he has made his own, but so long as the cubs, or pups, remain within it, he is their zealous and bold protector. The writer last quoted observes:

"It is surprising to see how few of the pups get crushed to death while the ponderous males or 'bulls' are floundering over each other when engaged in fighting. I have seen two bulls dash at each other with all the energy of furious rage, meeting right in the midst of a group of forty or fifty pups, trampling over them with their crushing weights, and bowling them over right and left in every direction, without injuring a single one. I do not think that more than one per cent. of the pups born each season are lost in this manner on the rookeries.

"To test the vitality of these little animals, I kept one in the house to ascertain how long it could live without nursing, having taken it immediately after birth and before it could get any taste of its mother's milk; it lived nine days, and in the whole time half of every day was spent in floundering about over the floor, accompanying the movement with a persistent hoarse bleating. This experiment certainly shows wonderful vitality, and is worthy of an animal that can live four months without food or water, and preserve enough of its latent strength and vigour at the end of that time to go far off to sea, and return as fat and hearty as ever during the next season."

It is often supposed that the sealskin we find at the furrier's is the animal's fur in a natural condition. Such is, however, by no means the case, for the freshly taken skin is not at all handsome looking. The beautiful fur which ladies know so well, is, in the natural condition, entirely concealed by a coat of stiff-brown or grey over-hair, which has to be carefully removed, and the skin is treated in a variety of ways before it is ready for the market.

It appears that about a million of sea-lions are born annually in the Prybiloff Islands, and that with the arrangements now happily effected, we need no longer fear their extermination, in spite of the prodigious number annually destroyed. Besides man, these animals have to dread sharks, sword-fishes, and above all the grampus.

The habits of the largest variety, Steller's sea-lion, are substantially like those of the fur-bearing kind, but it seems to be a less acutely jealous husband, and the herds do not form so many rows inwards from the shore. Its voice is a bark or a grand, deep roar.

Such are the creatures known as sea-lions and sea-bears, but to understand them fully we must endeavour to estimate the position in which they stand to other animals, which are their close, or their moderately distant, zoölogical allies.

Almost every one knows, to a certain extent, what a seal is. Of such animals there are said to be at least upwards of a dozen and a half distinct species, and nine are found in North America. Five of these nine are also to be met with in the northern part of the Old World.

Seals may at once be distinguished from sea-lions and sea-bears by the fact that they have no external ear whatever, so that they are known by contrast with the latter as the "earless-seals." None of them have soft woolly under fur like that of the furry sea-lions; but they have, what the "eared-seals" have not, namely, five well developed claws to each foot. Of the five digits of the hand, the thumb is slightly the longest, while in the hind foot the digits which answer to our great and little toes are the longest and the middle one is the shortest. The toes are webbed, but the hind-limbs are very differently disposed from those of the sea-lion. Every one who has observed a seal's progress on land must have been struck with the singularly awkward and wriggling movement of its body. This is due to the fact that though the seal is a quadruped as regards the number of its limbs, it is no quadruped as regards their use. The sea-lion, as before stated, does walk on all fours—in spite of the little freedom of the hind limbs— and does turn its hind feet forwards, resting on the soles of those feet as it walks.

The seal, however, is quite unable to turn its hind feet forward at all. Their soles are hairy, and it can never walk on them. On the land, therefore, the seal can only progress by contracting the muscles of its body, and by thus contorting its form, it is able to wriggle over the ground with more or less assistance from the fore-limbs. Nevertheless, it can thus shuffle along, especially over the ice, with surprising speed. Its hind limbs are only useful to it for progression when in the water, and then they are extended backwards, applied together—with sole to sole—and flapped, alternately right and left, so as to serve in the same way as does the tail of a fish.

Not only are the legs thus permanently bent backwards, but they are bound together by a fold of skin which also embraces, and is attached to, the sides of the short tail. Little therefore as may be the resemblance or affinity existing between the seal and the bat, there is a certain similarity in the construction of this region of the body in these two very different kinds of beasts. We saw, when considering the structure of the bat,* that

* See p. 153.

## THE SEA-LION

most of them have what is called an "interfemoral membrane," that is to say, a membrane which extends inwards on either side from the leg, and embraces the tail. The seal therefore may also be said to possess an "interfemoral membrane," though the influence it exerts is exercised in aquatic and not aërial locomotion.

Seals would be exceedingly numerous but for their constant destruction by man. What this must be is indicated by the fact that two hundred thousand individuals of the kind known as "the Greenland seal," are annually killed around Jan Mayen Island in the North

FIG. 75.

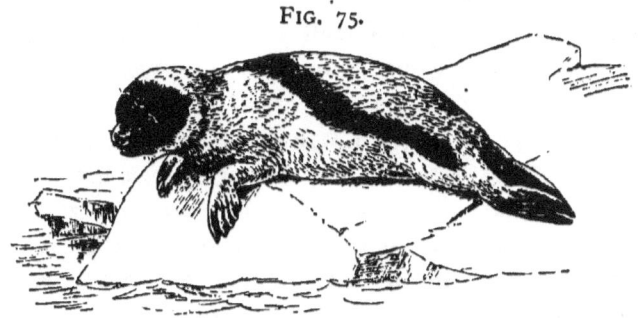

THE GREENLAND SEAL.

Sea, by the crews of Scotch, Dutch and Norwegian vessels. Seals—that is "true seals" as distinguished from sea-lions—are naturally inhabitants of all the shores of the temperate and colder regions, and one kind found north of the Equator is also found south of it, yet, with one exception, not even a single genus is common to the northern and southern hemispheres.

Many kinds of seals are gregarious, but others are solitary, and all are harmless animals to man if not attacked. They are very fond of basking in the sunshine, and spend a large part of their time on sand-bars, rocks, or on the ice, according to circumstances. They are very inquisitive animals, and most kinds are strongly attracted by musical sounds. Seals live mainly on fishes, but also on molluscs and creatures of the crab and lobster kind. They can remain a long time beneath the surface of the water, certainly for a quarter of an hour.

The young seals, strange to say, enter the water unwillingly, and have to be taught to swim by their parents. Some species will remain out of the water for three weeks after birth. Seals are remarkable for the affection they show to their young, and they are also very intelligent animals and readily learn to perform a number of surprising tricks.

Like the sea-lions, seals fall victims to sharks and to the grampus, but many are also destroyed by polar bears.

Several species migrate with regularity, and Mr. J. C. Stevenson has related how, during the summer and autumn, numbers of these creatures are met with in regions whence the approach of severe weather forces them to retreat southwards. This movement is anxiously watched for by the human inhabitants of the coasts along which they travel, and watchmen are set to communicate the news of the approach of the seals. They come at first in small detachments of from half a dozen to a score, and such will gradually increase in frequency for two or three days, when they come in hundreds. The main body then follows and averages two days in passing any spot. In all quarters, as far as the eye can carry, nothing is visible but seals, and the sea seems full of their heads. Iu about a week the whole host, consisting of many hundreds of thousands, will follow the polar current which sets through Hudson's Bay, and sweeps the coast of Labrador in a south-east direction. Then some go towards the Gulf of St. Lawrence, but most continue on till they come to the Gulf Stream on the banks of Newfoundland, which they reach about the end of each year. At the end of January they again begin to turn northwards.

The vast destruction of seals by man has greatly diminished their numbers in many localities, and has actually exterminated them in not a few. The common seal is found both in Europe and America, and on the Pacific as well as on the Atlantic sides of the latter continent. It is frequently met with in France and England, and is not rare in Spain. It is the only one which is common on the eastern coast of the United States.

This kind, like other species of the group, is certainly attracted by musical sounds; probably only through curiosity, because it is similarly attracted by any unusual movements. Mr. Bell tells us, in his "British Quadrupeds," that, in the Orkney Islands, if people are passing in boats, seals will often come quite close up to the boat, and stare at them, following for a long time together; if people speak loud, they seem to wonder what may be the matter! The Church of Hoy, in Orkney, is situated near a small sandy bay, much frequented by these creatures, and it was observed that when the bell rang for divine service, all the seals within hearing swam directly for the shore, and kept looking about them, as if surprised rather than frightened, and this continued as long as the bells rang.

Although it feeds mainly on fish, it will occasionally capture sea-birds, swimming beneath them and seizing them as they rest on the surface of the water, and they often make raids upon fishermen's nets.

The Greenland, or harp seal is (Fig. 75) another northern form common to both hemispheres. It is of a yellowish white colour, and the male has a crescentic black mark encircling the greater part of its back.

T

Individuals of this species are often extremely abundant on the shores of Newfoundland, where hundreds of thousands of them are annually killed in the spring. It has been often said to visit England, and it is certain that one was captured in Morecambe Bay in 1874.

At breeding time the females take up their stations on the ice very near each other, sometimes not a yard apart. The males accompany them, but mostly remain in the water, into which element the young, as before said, do not seem to enter voluntarily.

Professor Jukes* tells us of a young one which was taken alive and became a very gentle and interesting pet. "He lay very quiet on deck, opening and closing his curious nostrils—and occasionally lifting his fine lustrous eyes. On being patted on the head he drew it in till his face was perpendicular with his body, knitted his brows and closed his eyes and nostrils, thereby assuming a very comical expression of countenance. Although he was fierce when teased, and attempted to bite and scratch, he immediately became quiet on being stroked and petted."

The ringed seal is an Arctic species, which descends southwards in both hemispheres, and has been captured in England on the coast of Norfolk.

The Caspian seal, which inhabits the Caspian and the Aral Seas, seems to be very nearly related to the ringed seal, as also does the species known as the Siberian seal, which inhabits the lakes of Laikal and Orok.

The bearded seal, or, as it is called in England, the great seal, seems first to have been distinctly recognised in 1743, when one was shown at Charing Cross. It extends from the Arctic Seas downwards to both the North Atlantic and the North Pacific. It has a curious habit of turning a somersault when about to

* See " Excursion in Newfoundland," vol. i.

dive, by which it may be distinguished from other seals. Its food consists almost entirely of molluscs and creatures of the crab and lobster kind. It is said to be easily killed in the sea on account of its tendency to approach any boat; but on the ice it is very watchful. Its flesh is also reported to be more delicate in taste than that of other species.

The grey seal is exclusively confined to both sides of the North Atlantic. Its food consists mainly of fish, and especially of the tunny. In the beginning of October they seek rocks and islands which have not too precipitous shores, and which are not covered by the spring tides. There the females bring forth their young about the middle of the month, and these do not enter the water till they are four or five weeks old. During that time the young are lying upon the dry land they do not leave their places, but every tide their mothers crawl up to them to suckle them.

The bladder-nosed, or hooded, seal is distinguished, at least in the male sex, by the possession of a curious distensible muscular bag on the top of the head, extending backwards from the muzzle to some inches behind the eyes. Its "bladder" is altogether about a foot long, and when fully distended is nine inches high. This animal is restricted to the colder parts of the North Atlantic and to portions of the Arctic Sea, ranging from Greenland to Spitzbergen, but being rarely found south of Norway and Newfoundland. They are not common animals or so easily seen as some other kinds, since they swim low, with only the top of the head above the surface. The males fight fiercely, but when the various families are constituted, their affection for each other, and especially for their young, is said to be very strong. Both parents will remain so persistently with their pups, that the whole family are easily destroyed, though the hood of the male affords such a protection to its owner as to render the animal hard to kill with any ordinary club.

The largest and most singular of the seals is the sea-elephant, which attains the length of nearly twenty feet. The nose of the adult male is prolonged into a short, tubular proboscis. This ordinarily hangs down, but can be dilated and prolonged under excitement. Its hind feet are devoid of nails. It was formerly abundant

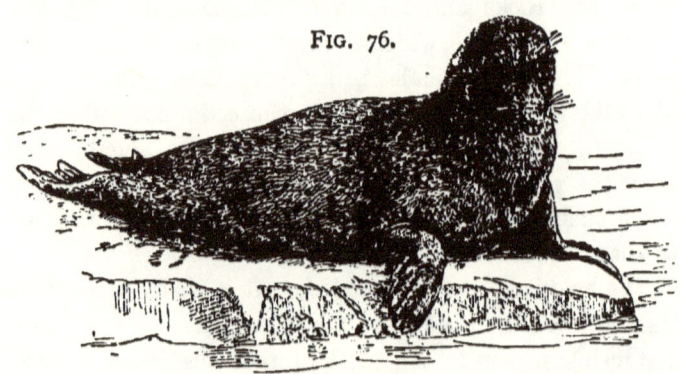

FIG. 76.

THE SEA-ELEPHANT.

in the Antarctic Seas, and also on the coast of California. There are two varieties, one confined to Western Mexico and Southern California, while the other is found in the Indian and South Pacific Oceans and the Antarctic Seas.

The former variety is the smaller one. It was formerly very abundant on the Mexican and Californian coasts, but has become nearly extirpated. Captain Seaman, who in 1852 had command of a sealing brig, has given the following account of the chase of these animals:

"The sailors," he tells us, "get between the herd and the water; then raising all possible noise by shouting, and at the same time flourishing clubs, guns, and lances, the party advances slowly toward the rookery, when the animals will retreat, appearing in a state of great alarm. Occasionally an overgrown male will give battle, or attempt to escape, but a musket-ball through the brain despatches it, or some one checks its progress by thrusting a lance into the roof of its mouth, which causes it to settle on its haunches, when two men with heavy ashen clubs give the creature repeated blows about the head till it is stunned or killed. After securing those that are disposed to show resistance, the party rush to the main body. The onslaught creates such a panic among these peculiar creatures, that, losing all control of their actions, they climb, roll and tumble over each other, when prevented from further retreat by the projecting cliffs. We recollect in one instance, where sixty-five were captured, that several were found showing no signs of having been either clubbed or lanced, but were smothered by numbers of their kind heaped upon them. The whole flock, when attacked, manifest alarm by their peculiar roar, the sound of which, among the largest males, is nearly as loud as the lowing of an ox, but more prolonged in one strain, and accompanied by a rattling noise in the throat. The quantity of blood in this species of the seal tribe is supposed to be double that contained in an ox, in proportion to its size.

"After capture the flaying begins. First, with a large knife the skin is ripped along the upper side of the body the whole length, and then cut down as far as practicable, without rolling it over. Then the coating of fat that lies between the skin and flesh—which may be from one to seven inches in thickness, according to the size and condition of the animal—is cut into "horse-pieces," about eight inches wide, and twelve to fifteen long, and a puncture is made in each piece sufficiently large to pass a rope through. After cleansing the upper portion of the body, it is rolled over, and cut all around as above described. Then the 'horse pieces' are strung on a rafter-rope and taken to the edge of the surf; a long line is made fast to it, the end of which is thrown to a boat lying just outside the breakers; they are then hauled through the rollers and towed to the vessel, when the oil is extracted by boiling the blubber. . . . The oil produced is superior to whale oil for lubricating purposes. Owing to the continual pursuit of the animals, they have become nearly, if not quite, extinct on the Californian coast, and the few remaining have fled to some unknown point for security."

FIG. 77.

THE WALRUS.

Our readers may gather from the foregoing sketch some notion of the group of seals, and so be enabled the better to understand, by contrast, the nature of the sea-lion. But there is an intermediate group which consists of the one very singular form known as the walrus or morse. Of these animals there are two varieties, one of which is found in the North Pacific and the other in the North Atlantic. Many naturalists, however, regard both these as constituting but one species.

The walrus differs from the seals in that the hind feet are naked and turned forwards in walking on land, though not so completely as in the sea-lions. It differs, on the contrary, from the sea-lions by not having any external ears.

But it diverges from both these groups in that its teeth are all small, simple, one-rooted, and with flat crowns except the two upper eye-teeth, which are developed into immense tusks that descend a long distance below the lower jaw. The head is round, with a very short and broad muzzle furnished with stiff bristles. The eyes are rather small. The tail is very rudimentary. The fore-feet have toes of nearly equal length, each with a minute flattened nail. The hind feet have the fifth toe slightly the longest, and it and the great toe bear minute flattened nails. The nails of the other three toes are long and pointed. The toes of the hind foot are furnished at their ends with processes of skin as in the sea-lions.

The walrus is a very heavy bulky animal, which is especially thick about the shoulders, and measures eleven feet from the snout to the end of the tail. The body is covered with short, yellowish-brown hair, but this may become very scanty or almost entirely disappear with age.

One of the earlier accounts of this animal relates how some two hundred walruses were met with by William Barents, a Dutch navigator, in 1594, lying on the shore of Oray Island. It was, however, mentioned by Albertus Magnus in the 13th century, and it was figured in 1568.

The tusks, which are stronger in the male than in the female, are formidable weapons of about a foot and a half in length when fully developed, if not two or three inches longer. They are most powerful means of defence against Polar bears, or any other enemies which can be reached at close quarters. It is often said that they are used in climbing, but they seem to be principally employed to scrape and dig amidst the sand and shingle for the molluscs and other "shell-fish," which constitute the principal food of the walrus. It also feeds on sand-worms, starfishes, and shrimps. Various kinds of seaweed have been found in its stomach, though it is not certain that such vegetable substance was intentionally swallowed.

It extends as far North as explorers have yet gone, and on the land round Hudson's Bay, Davis' Straits, and Greenland—but in rapidly decreasing numbers. It still frequents Spitzbergen, Nova Zembla, and the western part of the north coast of Siberia, as also northern Kamtschatka, Alaska, and the Prybiloff islands. Fossil remains show that it once inhabited France, Belgium, the United States, and England, and it has occasionally visited the British Isles during the present century.

The word "walrus" is a modification of the Scandinavian name "whale-horse." "Morse" is from the Russian "morss." The Lapp word is "morsk."

Few animals have been more thoroughly misrepresented in figures, than has the walrus. The reader interested to see copies of these, or who desires to be furnished with full details concerning all the animals here noticed, is referred to an admirable work on North American Pinnipeds, by J. A. Allen, and published at the Government Printing Office, Washington, in 1880. One of these strange figures represents the walrus with a fish-like body, covered with scales, a pig's head, tusks directed upwards, and long ears. Others, hardly less monstrous, also depict the animal with ascending tusks; but in 1613, an admirably correct figure of the creature was given by Hessel Gerard, in which the hind feet are represented as being turned forwards. It is curious that later figures, until quite recent times, were made to depict the animal with its hind limbs turned backwards like those of a seal.

The walruses are, according to Allen, always more or less gregarious, in larger or smaller companies according to their abundance. They seem to delight in huddling together on the ice-floes, or on shore, to which places they resort to bask in the sun. They are rarely seen far out in the open sea, and are restricted in their wanderings to the neighbourhood of shores, or large masses of floating ice. Though they often move from one feeding-ground to another, they do not truly migrate. Usually, but one young one is born at a time, never more than two; and this takes place between April and June. The females (with their young, which they have been said to suckle for two years) seem to consort together. Of thirty full-grown walruses killed in Henlopen Straits, in the month of July, not one was found to be a male. On some other occasions, however, males, females, and young have been found together. Their strong affection for their young, and their sympathy for each other in times of danger, have been repeatedly noticed.

Mr. Lamont, in his "Seasons with the Sea-horses," says: " I never witnessed anything more interesting and more affecting than the wonderful maternal affection of the walrus. I perceived that she held a very young calf under her right arm, as she saw that Christian wanted to harpoon it; but whenever he poised the weapon to throw, the old cow seemed to watch the direction of it, and interposed her own body, and she seemed to receive with pleasure several harpoons which were intended for the young one."

When one individual has been wounded, the whole

herd act altogether in common defence. If not interfered with, they are harmless and inoffensive, though they are fierce enough if attacked, when they prove dangerous antagonists. Owing either to confidence in its own powers, or want of appreciation of the danger of human foes, it has been said to be, as a rule, not easily alarmed, but permitting a near approach before manifesting uneasiness or fear. Other accounts, however, describe walruses as wary animals, usually keeping a sentinel on guard while the herd is asleep.

Their voice is a loud roaring, and can be heard at a great distance. Dr. Kane has described it as "something between the mooing of a cow and the deepest baying of a mastiff, very round and full, with its bark or detached notes repeated rather quickly seven or nine times in succession.

As to the cruel and useless slaughter of these animals, Lamont tells us that in August 1852, two small sloops sailing in company approached an island, and soon discovered a herd of walruses, numbering, as they calculated, from three to four thousand, reposing upon it. Four boats' crews, or sixteen men, proceeded to the attack with spears. The great mass of walruses lay in a small sandy bay, with rocks enclosing it on each side. A great many hundreds lay on other parts of the island at a little distance. The boats landed a little way off, so as not to frighten them, and the sixteen men, creeping along shore, got between the sea and the bay full of walruses. The walrus is very active and fierce in the water, when a herd will keep wonderfully together as they dive and reappear, a hundred grisly heads, with long gleaming white tusks, appearing above the waves at the same moment. On shore, however, they are very unwieldy and helpless, and those in the front soon succumbed to the lances of their assailants. The passage to the shore soon got so blocked up with the dead and dying that the unfortunate wretches behind could not pass over, and were in a manner barricaded by a wall of carcases. Considering that every thrust of a lance was worth twenty dollars, the scene must have been one of terrific excitement to the men. They slew, stabbed, slaughtered, butchered, and murdered, until most of their lances were rendered useless, and themselves were drenched with blood and exhausted with fatigue. They next went on board their vessels, ground their lances, and had their dinners, and then they returned to their sanguinary work; nor did they cry "hold, enough!" till they had killed nine hundred walruses, and yet so fearless and so lethargic were the animals, that many hundreds more remained sluggishly lying on other parts of the island at no great distance.

A walrus was brought to London in the reign of James I. "When the King and many honourable personages beheld it with admiration for the strangeness of the same."* Two specimens have of late reached the London Zoölogical Gardens, but lived but a very short time. That this species possesses docility and intelligence similar to that of the seal, is shown by some observations reported by Mr. Brown † with respect to a young one he saw on board a ship in Davis' Straits in 1861, and which had been caught off the coasts of Greenland:

"It was fed on oatmeal and water and pea soup, and seemed to thrive. No fish could be got for it, and the only animal food which it obtained was a little freshened

* "Purchas, his pilgrimes," 1624, vol. iii. p. 560.
† In a very interesting communication made to the Zoölogical Society of London on June 25, 1868.

beef or pork, or bear's flesh, which it readily ate. It had its likes and dislikes, and its favourites on board whom it instantly recognised. It became exceedingly irritated if a newspaper was shaken in its face, when it would run open-mouthed all over the deck after the perpetrator. Sometimes it would run at a clumsy rate into the surgeon's or captain's cabins, or from one side of the ship towards the other and back again, in imitation of such a movement ('sallying') on the part of the sailors. It lay during the day basking in the sun, lazily tossing its flippers in the air, and appearing perfectly at home. One day the captain tried it in the water for the first time; but it was quite awkward and got into the floe, where it was unable to extricate itself till its master went out on the ice and called it by name, when it immediately came out from under the ice and was, to its great joy, safely assisted on board again, apparently heartily sick of its mother element. It lived a little over three months."

Mr. Lamont captured several young walruses; three of them were kept in a pen on board ship together. One of them "Tommy," was a great pet, but to the general grief he was one day found dead, with his face immersed in a pail of gruel, and one of the others lying on the top of him—clearly suffocated.

On reviewing the facts herein stated, it will be seen that the sea-lion belongs to a group of aquatic four-legged beasts which is divisible into three groups: (1) eared seals, (2) walruses, and (3) true seals, of the first of which three it is a member. The whole group is considered to rank as an "order," known from the peculiar modification of their paws, as the order of Pinnipedia.

But to what other order of beasts are the Pinnipeds allied? There can be no doubt that they are allied to the Carnivora, or beasts of prey, and not at all to the porpoises and dolphins—though these latter are also aquatic, warm-blooded beasts, and not fishes. Nevertheless, the organisation of the Pinnipeds is in many respects very different from that of the ordinary Carnivores, so that there might be a suspicion that such resemblances as do exist between them may have been due, as we have seen to be the case in so many other cases, to an independent origin of similar structures.

But a certain curious and recondite similarity of internal structure indicates the existence of a real affinity between Pinnipeds and Carnivores. In the latter—in dogs, cats, civets, bears, weasels, and racoons—some folds of brain substance, on the anterior portion of the upper surface of the brain, give rise to a certain pattern and appearance which may be compared with what is known in heraldry as an "escutcheon of pretence." A careful examination of the brains of seals, walruses, and sea-lions, shows that they have also the same condition of brain structure though it is not at first so readily apparent as in ordinary Carnivores. Since such a resemblance can hardly be the result of surrounding conditions and so have arisen independently, we think it may be safely regarded as a true indication of the existence of a real, more or less hidden, affinity. But though we thus seem forced to admit a genetic relationship between Pinnipeds, and, at least, the ancestors of our existing terrestrial Carnivores, it does not follow that all Pinnipeds have had the same origin. There are many resemblances between the sea-lions and ordinary bears, and one such resemblance consists in the fact that the members of both these groups possess in the skull that canal for a branch of the carotid artery which we have, (in our article on the American bison), called the "canal in the wing-wedge bone of the skull." Such a canal is wanting in ordinary seals, and it is wanting also in that aquatic modification of the weasel type of structure, the otter. It is possible, therefore, that sea-lions and bears may have had one common ancestor, and seals and otters another and distinct common ancestor. If so, much of that resemblance between seals and sea-lions which is related to their similar aquatic habits of life, may really have arisen independently, so that they together form, in our own day, a much more homogeneous group than they formed at some anterior epoch. This, however, is mere speculation, and we are far from wishing to insist upon even its probable truth. It is an interesting possibility and no more.

The argument against it is that bears seem to be a very modern group of mammals, and it may be said that both seals and sea-lions are descendants, not of any forms which closely resemble existing land-Carnivora, but rather of certain beasts, remains of which have been found deep down in tertiary strata—in the Eocene formations of Europe and North America—and which are known to naturalists as the "Creodonta." Such creatures have been closely studied in America by Professor Cope, and it may turn out not only that they were the common ancestors of the existing Pinnipeds and land Carnivores, but also of the whole of those much more divergent groups, which together constitute the ordinary or placental mammals of to-day, together with the pouched beasts which are distinguished from them as Didelphous mammals,* as was explained at some length in our article on the opossum.

* See p. 62.

## XI

## WHALES AND MERMAIDS

IN the time of Alexander the Great and afterwards under the Seleucidæ, the ancient Greeks became acquainted with the north-western part of India. Then and there they heard many strange tales, which, as usual (especially when two different races and languages are concerned), lost nothing in the telling. Among other things, they heard that the seas about Ceylon were peopled with mermaids. In this case, as in the case of so many other wonderful tales, there was a certain amount of truth underlying the fiction; for those seas are peopled by creatures (as big or bigger than human beings), which have a habit of raising themselves up vertically out of the water, when they present a very startling appearance to an unscientifically critical eye. Astonished travellers beheld beings with rounded, human-looking heads, showing their body down to the bust out of the water, displaying a pair of rounded prominent breasts, and not seldom holding a baby in their arms. After remaining some time in this attitude, they would suddenly dive, and then a tail like a fish's became exposed to view. Small wonder, then, that sailors should imagine they were beholding creatures half woman and half fish, for the vivacity of a sailor's imagination is proverbial.

But the creature thus seen is as different in temper and habits from the fabled mermaid as it is in body.

Instead of seeking to seduce unwary voyagers to visit its home beneath the waves, in order there to devour them, the dugong (for that is the name of this sort of mermaid) browses peacefully on seaweed, and is as harmless as it is curious.

It is a creature which, as ordinarily met with, is about eight feet long. Only a faintly marked neck is visible between the head and the trunk, which tapers gradually backward to end in a horizontally flattened tail. Unlike the seals and sea bears, the dugong has no trace of any hind limb, and has only a pair of short paddle-shaped fore-limbs, the five digits of which are enclosed

Fig. 78.

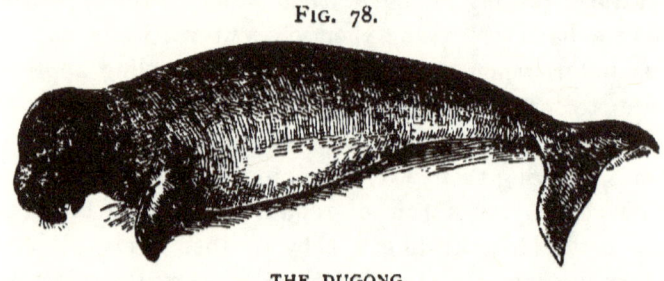

THE DUGONG.

in a common fold of skin, and are not therefore visible externally. They have no nails. Deep in the body of the animal are small bones which are the rudiments of that bony structure (called the pelvis) to which our thigh bones are articulated. But there is no rudiment representing the thigh bone itself.

The skin of the body is very thick, rough, and almost naked, but with a few hairs. Some hairs extend inside the cheek, and there are stray bristles on the lips. The eyes are small, there are no external ears, and the nostrils can be closed, having each a valvular external aperture.

We have met with, in the otters, animals specially organised for an aquatic life, and in the sea-bears and especially in the seals, creatures yet more exclusively so constructed, since the last-named animals can progress on land only with awkwardness and difficulty. Still all these beasts can so progress, either in quadrupedal fashion —as do otters and sea-bears—or by convulsive bodily contortions, as do the seals. But in the dugong, for the first time (in our survey of different forms of life) we come upon a creature absolutely aquatic and quite unable to live on land. Indeed, not only does it remain afloat, but it even avoids very shallow water, partly on account of its terrestrial helplessness, and partly on account of its seaweed diet.

It is found in the Red Sea, off the east coast of Africa, near Ceylon, in the islands of the Bay of Bengal, and the Indian Archipelago, including the Philippine Islands, and on the north of Australia. Thus it may be said to range the Indian Ocean and a portion of the Pacific.

In Australia the dugong is now regularly "fished" on account of its oil, which is peculiarly clear, limpid, and free from any disagreeable odour, and is said to have the same salutary qualities as cod-liver oil. It is a slow, inactive, mild, and inoffensive animal, incapable of self-defence, and apparently destined ere long to become extinct and disappear, as we shall see shortly that one of its near relations has already done.

Before passing to the nearest surviving species, a word or two must be said as to its teeth and the structure of its palate. In the first place, the male dugong possesses a pair of large, nearly straight tusks, which project downward to a short distance beyond the mouth. They may remind the reader of the tusks of the walrus, but they are shorter and of a different nature, for they are

U

not "canines," but "incisors"; that is to say, they do not answer to our "eye teeth," but to two of our "upper cutting teeth," which are placed between our eye teeth. The creature has some grinding teeth, but what is most curious is the presence of a large, rough, horny plate which clothes the front part of the palate, and another similar plate which rubs against the former, and clothes the front of the lower jaw. The reader may perhaps recollect that in ruminants there is a small horny pad at the front of the upper jaw, against which the teeth of the lower jaw bite. This pad, however, is only a

FIG. 79.

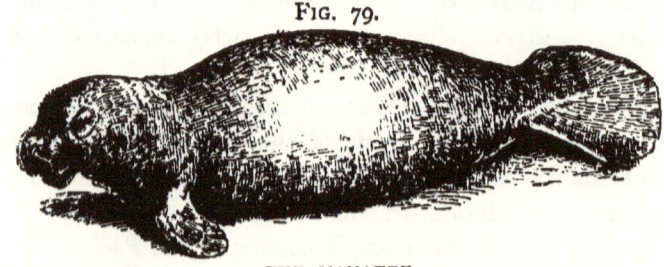

THE MANATEE.

mere rudiment of that we meet with in the dugong. But, in our survey of the creatures noticed in this article, we shall shortly describe a very much exaggerated structure of a more or less similar kind.

The manatee is another "mermaid," and a cousin of the dugong, which it closely resembles in general form. It is a denizen of America, and even of the United States, as it is found in Florida as well as in some of the West Indian Islands, and in South America to 20° South latitude. It ascends high up in the rivers of Brazil, and is found on the west coast of tropical Africa and also in its rivers, even as far into the interior as Lade Tchad.

Its length does not appear to exceed eight feet. It has, like the dugong, horny plates in the front of its jaws, but differs from the last-named animal in having no tusks, though it has more grinding teeth. In the young there are rudimentary teeth concealed beneath the horny plates. As they never penetrate these plates they must be useless, and they quite disappear before the animal is adult. The manatee has a very peculiar upper lip, which has a median division, on each side of which is a lateral lobe or pad. These pads may either be moved apart or be brought closely together, and thus the animal can grasp its food. When about to feed it will first separate the two lateral lobes and then close them upon the branch or leaf it is going to feed on, afterward bending back the whole lip, so as to introduce the food thus seized into its mouth without any need of employing the lower lip for this purpose.

We saw when studying the sloth that the number of bones in the neck of nearly all beasts is, as also in man, seven. Such is also the case in the dugong, but in the manatee—though its neck is no shorter—there are but six such bones.

The name "manatee" seems to have been given originally to this animal by some of the first Spanish settlers in the West Indies, on account of the strangely free and hand-like use it can make of its paddle-like fore limbs. It uses them for bringing food towards its mouth, and can bend the wrist and elbow, as well as the shoulder-joint. There are generally also more or less rudimentary nails on the fingers.

The manatee differs in habits from the dugong in that it frequents rivers, estuaries, and lagoons, preferring shallow water, and quite eschewing the open sea. It feeds exclusively on aquatic plants, on which it browses under water, and is extremely slow in its movements and inactive. It has a small and simply formed brain, and is harmless and perfectly inoffensive. In deep water it often floats with its body much arched, its head and tail hanging downward. In shallow water it will support itself on the end of its tail or will crawl about, only applying the tips of its paddles to the ground.

Manatees never voluntarily quit the water, not only on account of their being so extremely unsuited for progression on land, but also because (from the structure of their bodies) they cannot there breathe at their ease.

In 1878 a fully grown female was caught in British Guiana, where they now seem to be getting very scarce. On the voyage across the Atlantic it was kept in a large box two-thirds filled with fresh water. This was placed near the donkey engine, so that steam could every now and then be passed into the water to maintain the temperature of the latter at a steady warmth in colder latitudes. Having arrived at Greenock, it was conveyed to London by rail, warm water being occasionally poured into its tank on the journey. During the night the manatee frequently raised itself and tried to get out of its box. After its arrival at the Westminster Aquarium it was nearly a week before it would feed. Its owners, alarmed for its life, then fed it by force. The water was drained off from its tank, and three persons entering it, inserted a cork in the forepart of the mouth, whereupon some milk was injected by a syringe. The manatee, though ordinarily exceedingly quiet and gentle in its demeanour, evidently objected much to the proceeding, and, though obliged to swallow some of the milk, rejected what it could, using so much force that it was all the three men could do to restrain it. But neither then nor at any other time did it utter a sound, nor attempt to bite or in any other way injure its assailants, though floundering, wriggling, and struggling with all its might. Thenceforward it fed spontaneously on the green food given it which floated in the water of the tank. Its favourite food was lettuce, but it would also eat cabbage and watercress, and altogether consumed from 90 to 112 pounds of green food daily. Its tank was kept at a temperature of from 70° to 74° Fahrenheit, and for six months all went well. But unfortunately, about Christmas, during very cold weather, its keeper accidentally allowed the water one night to drain away, so that it was left dry in a cold atmosphere. Next morning, after being freshly supplied with water, it appeared ill. It refused food, and became thinner and thinner, till it died from exhaustion on the 15th of March 1879.

Quiet stolidity and stupidity seem to characterise it in its native haunts as well as in captivity. The Aquarium specimen was nocturnal in its habits, feeding by night. During the greater part of the day it dozed in various attitudes, every now and then rising lazily, and apparently without the slightest effort, to the surface to breathe; or occasionally it made a move round the tank in a quiet, unconcerned manner. Then it would poke its nose close up to the glass, remaining stationary there for a time without showing either fear of, or interest in, the numerous spectators frequenting the Aquarium.

A fine, robust young male arrived at Liverpool from Trinidad in September of the same year, and was purchased for the Aquarium at Brighton, where it was kept with a young female that was obtained a few months previously. They are said to have recognised the voice of their keeper, and seemed to enjoy having their backs brushed by him. It is reported* that they habitually

* " Proceedings of the Zool. Soc. 1881," p. 456.

assumed a horizontal position, the body, when resting on the ground, being supported by the under surface of the tail fin, and it may be that the posture assumed by the Westminster specimen was due to one of its paddles having been injured. They eat by preference lettuces and endives, and these were always swallowed under water, and they never eat when removed from it, though food was repeatedly then offered them. When out of the water they seemed to be oppressed with their own bulk, and could only progress a few inches by means of pressing their jaws and tail fin closely to the ground, and making violent lateral efforts of the body, slightly supported by the paddles.

The male devoured his food more rapidly than the female, and thus obtained an undue share, so that it was thought advisable to separate them at feeding time. For this purpose a wooden partition, fitting into a groove in the floor and fastened by upright supports, was occasionally let down into the tank, projecting a few inches above the surface of the water. The female took no notice of this alteration, but invariably waited before commencing to feed until her mate was supplied on his side with a portion. The necessity for the separation soon became apparent; for the male cleared up every scrap of food long before the female, a more dainty and delicate feeder, had finished. He then became very restive, swimming actively around his straitened quarters, pressed his nose against the partition, rolled over on his back, and exerted considerable force in his obstinate and repeated attempts to remove the obnoxious obstacle. Failing in his endeavours to push it on one side, he next tried to get over it, lifted his head above the water, feeling the edge of the partition with his fore paddles and raising them till they were almost level with the projecting edge.

In the spring of the year 1880, the female manatee died, after several months' existence in the Aquarium. The history of the male in the subsequent interval may be epitomised in the words "lie still and grow fat." He evinced no grief at the loss of his companion.

The dugong and the manatee are the only two mermaid kinds now existing on the surface of this planet. But a little more than a hundred and twenty years ago there was a third kind, much larger than either of the existing ones, as it attained a length of from twenty to twenty-four feet. It was the *Rhytina*, and its destruction is one of the few well-attested examples of the extirpation of a species altogether by human agency. When first found it abounded, but very soon it entirely disappeared.

Eastern Siberia was not known to Europeans before the seventeenth century, but in the latter part of it that region came into the possession of Russia, after which it was visited by hunters and peopled by emigrants, who hunted the fur-bearing animals to be found there.

In 1718 Peter the Great sent a special mission to explore the chain of the Kurile Islands, and a little later, in 1727-29, another expedition set out, under Behring, thoroughly to explore Kamtschatka. Behring returned and made his report, but no such animal as the rhytina is mentioned in it. Some years later, in 1740, Behring visited Kamtschatka again and spent the winter there, having with him the remarkable and energetic naturalist, Steller, too early lost to science. Nevertheless they did not find the rhytina, and no one else has ever found it there, though large rewards have been offered for its discovery in that country. In 1741, however, Behring went again to the eastern shore of Asia, when he fitted out two ships, in one of which certain individuals embarked, Behring and Steller being of the company. They then crossed the North Pacific, and having (for the sake of a reward) stopped a few hours on the American coast, sought to return as quickly as possible. They were, however, wrecked on a little island at no great distance from the coast of Kamtschatka, now known as Behring Island. There Steller met with this animal, afterwards named rhytina by the naturalist Illiger. Steller, who was on the look-out for American things, took the animal to be that American mermaid, the manatee. Probably because of this and on account of the enormous multitude of individuals met with—perhaps also for lack of space—he took no part of the animal back with him. But he found that the creature's flesh was very good to eat, and so recommended traders to use it for provision. This advice was only too readily and perseveringly acted on, for in twenty-seven years from that date not a single living rhytina remained, the last being killed in 1768, so far as any certain information has been obtained. It appears never to have inhabited the Aleutian Isles, nor America, nor Kamtschatka, nor the Kurile Islands, but when first discovered was extremely numerous at Behring Island, finding abundant food in the large seaweeds which float about the coast. But its habits and disposition easily account for its rapid destruction. Like the manatee, the rhytina was very voracious, but it only fed in shallow water, and had very frequently to come to the surface to breathe. It was also exceedingly stupid and dull of sight and hearing, but perhaps its affectionate feelings were even more fatal to it, for if either a male or female were harpooned, its mate remained beside it and made endless stupid efforts to relieve it.

So completely destroyed was it that people became sceptical as to its ever having existed. But Brandt found in the Museum of St. Petersburg a horny plate which exactly resembled that which had been figured by Steller (and it was the only thing he had figured) in his account of the animal. The discovery of this plate thus served to prove both the truth of Steller's narrative and the animal's, previously unknown, nature. Afterward an imperfect skull was found at Behring Island, then three nearly complete skeletons were discovered, and recently yet other bones have been extracted from the frozen soil.

In form it resembled the dugong and manatee, but its head was relatively smaller and it had no teeth whatever, only a horny plate in each jaw. It had a thick, rugged, naked skin, though there were brush-like hairs on the paddles. It was of a dark brown colour, sometimes spotted or streaked with white.

The extinction of this animal may remind our readers of what was said in our notice of the turkey about the extirpation of the dodo. That bird had, like the rhytina, no means of escape or defence, was good for eating, and entirely confined to a minute and remote part of the earth's surface.

But how came the rhytina to dwell in such a tiny, out-of-the-way spot, and where did mermaids come from, and what may have been their ancestors? That their ancestors were quadrupeds and were once widely distributed over the earth's surface there can be no doubt. In the middle and later tertiary times mermaids of different kinds abounded in the European seas and swam about on the English coast where now is Suffolk. They were more or less like the dugong, but, though some of them were larger, their tusks were smaller. Their typical form has been named *Halitherium*, and the most remarkable thing about it is the fact that it had a pair of small thigh bones, though there could have been no external appearance of hind-limbs any more than in the three previously described mermaids. The naturalist Illiger, who gave the rhytina its name, called the small group I have spoken of as mermaids, by the name of Sirens, and the group (order) is now known by naturalists by the term *Sirenia*. The mode and source of their evolution are still quite unsolved problems, but there are not wanting indications that they may be collateral descendants of elephants and *Dinotheria*. If so, they can put in some claim to rank as odd-toed ungulates, absurd and paradoxical as it may seem to reckon among odd-toed hoofed beasts, creatures which have not only no " hoofs," but no " toes " either !

But if we cannot positively say what are the nearest relatives of the mermaids, our predecessors reckoned them as belonging to the group of whales and porpoises—an order termed by naturalists *Cetacea*. Our " mermaids " were formerly spoken of as the " Herbivorous Cetacea," to distinguish them from the creatures belonging to the other group of creatures (the whales and porpoises), all of which live on animal food.

To the consideration of these latter, which are the only true Cetacea, we will now turn.

They offer a most wonderful example of the puzzling and often misleading effects which external conditions can sometimes bring about, and are a notable warning how necessary it is when we seek to find out the affinities of the different animals not to rely much upon external characters when these are closely related to their mode of life. Whales and porpoises were long considered, very naturally, to be " fishes," and were classed among them even by the great naturalist, Ray. Their general form of body—which is spindle-shaped, with no sign of a neck

## WHALES AND MERMAIDS

between the head and the trunk, while posteriorly it tapers gradually to and in an expanded tail fin—is very fish-like, while their single pair of paddles are much more like fins than are those of the dugong and manatee. Nevertheless, in all essentials, whales and porpoises are true "beasts." They possess all the characteristics of that class, and are both warm-blooded and suckle their young. Not only is a whale much more like a bat or a squirrel than it is like a fish, but in many respects there is much more difference between a fish and a whale than there is between a whale and a humming-bird.

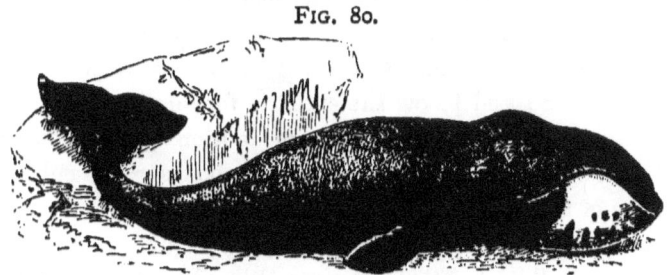

FIG. 80.

THE GREENLAND WHALE.

But whales and porpoises form a group or order of animals which is exceedingly well defined and distinct from every other order of mammals. Of all beasts they are the most completely and exclusively organised for aquatic life, being perfectly helpless on land, more so than even the dugong and manatee, and out of all comparison more so than seals or otters. On the other hand, no beasts are so perfectly at home in the open ocean, where the majority of species constantly disport themselves, though a few are inhabitants of rivers.

The true, or Greenland, whale is one of the largest animals which now lives, or, so far as we yet know, ever has lived, being from forty-five to fifty feet in length. More than one-third of this is occupied by its enormous head, the vast size of which is due to the great jaws which enclose what one might call an immense cavern containing the tongue and a quantity of horny plates—the so-called "whalebone." The upper jaw-bone is very narrow from side to side, but much arched from before backwards, while the lower jaw is greatly arched outward on either side. The upper lip is rudimentary, but it is met by a prodigious lower lip which stands up stiffly with a very convex margin from before backwards. Just behind the mouth is the small eye, close behind which again is the scarcely perceptible opening of the ear. The nose opens near the summit of the head by two crescentic apertures which can be opened or closed at will. A little behind and below the eye, the fore-limb, or paddle juts out. This has no power of motion except at the shoulder-joint, although inside it are bones representing those of the upper and forearm and of the five fingers of man and other pentadactyle beasts. But whereas in man and all such beasts the number of bones in every finger never exceeds three, here there are five in what represents the middle finger, and four in the skeleton of the digit on either side of it. At the hinder end of the body is a tail-fin in the form of two lateral pointed expansions of skin, supported by a dense fibrous substance within. Though no trace of any posterior limb is visible externally, there is, deep in the interior of the animal, a bone, only about eight inches long, which probably represents the thigh bone and bears at its extremity a small ossicle, which may be regarded as a rudiment of the shin bone. The former of these two bones is attached to a rudimentary representation of the pelvis, which exists here as well as in the mermaids. Although the neck is so short as to be imperceptible externally, there are the usual seven bones in it, though they all become united into one mass, or all of them save the seventh. The surface of the skin is smooth and glistening, and quite devoid of hair, but the body is kept warm by means of a thick layer of fat—the so-called "blubber" —which lies immediately beneath the skin. Within the enormous mouth there is, on either side, a series of long, flattened, horny plates (the whalebone), which grow on, and hang down from, the roof of the mouth. They thus form two longitudinal series, each plate of which is placed transversely to the long axis of the whale's body, and all are very close together. The outer edges of the plates are solid and nearly straight, but their inner edges incline outward, each plate becoming narrower as it extends downward. These oblique inner edges are also furnished with numerous coarse, hair-like processes, consisting of some of the constituent fibres of the horny plates, which as it were fray out, and the mouth is thus lined, except below, by a network of countless fibres projecting from the inner edges of the two series of plates. This network acts as a sort of sieve. When the whale feeds, it takes into its mouth a great gulp of water, which it drives out again with its tongue through the intervals of the horny plates of baleen, the fluid thus traversing the sieve of horny fibres which retains the small creatures—shrimp-like creatures and molluscs— on which these marine monsters subsist. Water in the mouth is no impediment to the whale's breathing, as the upper part of its windpipe (the larynx) passes up into and is enclosed by the back part of the nostrils, and thus no water can pass into the windpipe from the mouth. The longest of the baleen plates attains a length of ten or twelve feet, and there are some three hundred and eighty on either side, the series consisting, of course, of short plates at each end, the longest being in the middle of either series. They are so long that when the mouth is shut they lie back along its floor, their elasticity straightening them when the mouth is opened. It is to these horny plates that we referred when, in our notice of the dugong, we said that we should later describe a very exaggerated structure, somewhat similar to the palate plate of that animal. The adult whalebone whales are entirely devoid of teeth, though before birth many minute calcified teeth are formed in each jaw. But these are entirely absorbed and disappear before birth. The brain is four or five times as massive as that of any other animal.

The Greenland whale is known as the "right" whale because it is the right kind for the fishermen who seek for whalebone and blubber. It ranges round the North Pole, and is found on both sides of Greenland, and off the coast of Labrador. In Behring Sea and the Sea of Okhotsk its southern limit seems to be latitude $54°$. It is possible, but very improbable, that a straggler may have reached the British coast.

Much has been mistakenly said about the "blowing" and "spouting" of whales and other cetaceans. They do not really send out water from their nostrils, but only their breath when they breathe. They do not, of course, breathe rapidly as do land animals, since they require to come to the surface to do so. This is the reason why the tail is expanded horizontally in whales and mermaids. That shape helps them thus to rise by striking with the tail, while fishes, which do not need thus to rise, have the tail fin expanded vertically. When cetaceans rise to breathe, they forcibly expel a great volume of warm, moist air from their lungs. This ordinarily takes place in a cold atmosphere, and

## WHALES AND MERMAIDS

always does so in the Greenland whale, which is never far from ice. The column of warm moist air thus becomes immediately transformed into a cloud of minute particles of water. Besides this, when they "spout" before quite reaching the surface they may also raise up a jet of water, which their act of expiration displaces and casts upward.

Although a "right" whale never visits, and probably never did visit, the temperate part of the Atlantic, there is a southern kind—with a shorter head and less baleen—which is found in the temperate seas of both the northern and the southern hemispheres, and presents four varieties, often reckoned as species. One of these varieties inhabited the North Atlantic, and no doubt was often seen, in early days, "spouting" as it traversed the Straits of Dover. Four or five hundred years ago it was exceedingly common, and in the Middle Ages was keenly pursued by the Basques. From before the Norman Conquest, till the period of the Reformation, oil and whalebone were sent over Europe from Bayonne and San Sebastian, and from other places between those cities. As they grew scarce, the Greenland whale was met with in seeking a north-west passage to India, and has since become the great object of pursuit. Still, the southern kind has visited the Spanish coast so late as 1891, while in 1877, one came to southern Italy. It may now also be seen occasionally in New York Harbour, the Delaware River, and the coast of Maryland.

A whale known as the "humpback," so called because it possesses a dorsal fin (which the right whales do not) of a low hump-like form, ranges the Atlantic from Greenland and Norway, and sometimes makes its appearance on the coasts of the British Isles. Its length is from forty-five to fifty-five feet, and the female is the larger.

Its most remarkable character is the great length of its fin-like arms, and it differs also from the true whales in having numerous long grooves, or folds, extending beneath the throat.

Certain whales, known as rorquals, finbacks, or razorbacks, have still more numerous folds beneath the throat; they have also a dorsal fin, but only four fingers are enclosed in their relatively short limb. These are the commonest kinds of whales, and some of their varieties are to be found all over the world except in extreme polar regions. The common rorqual is the largest animal known, attaining sometimes a length of seventy feet. It feeds on fishes and largely on herrings, but other varieties feed exclusively on shrimp-like creatures.

The rorquals have much shorter whalebone and much less blubber than have the right whales, so that they were little cared for till of late, when on account of the increasing rarity of the more valuable species, rorquals have begun to be regularly fished. The grey rorqual frequents the western shores of the United States from December to March, and the females enter the lagoons of Lower California to bring forth their young. In October and November they skirt the coasts of California and Oregon going southward.

The toothed whales are far more numerous in species than are the whalebone whales. They ought rather to be called "whaleboneless" than toothed, as a few kinds have no teeth, while a whole section of the group is without any teeth in the upper jaw, and there may be but a pair in the lower jaw.

The sperm whale, or cachalot, is the giant of the group, attaining a length of from fifty-five to sixty feet. One-third of this total length is occupied by the head, which, seen in profile, has a rectangular anterior end, being truncated vertically in front. Unlike the right whales, the lower jaw is small (without any prominent upwardly projecting lip), is set with numerous simply conical teeth, and does not extend so far forward as the muzzle. The bones which support the immense upper jaw, do not by any means correspond with it in shape, for the upper surface of the skull is much lower and concave. The great mass of the upper jaw consists only of about a ton of an oily substance which yields "spermaceti," while the blubber, which everywhere copiously clothes the body, is the source of what is known as "sperm oil." The substance known as "ambergris" by perfumers, is also a product of this animal, being a concretion formed in its intestines. The nostrils have but a single external aperture, which opens close to the front end of the top of the snout, a little to the left of it, and so the animal "spouts" forward and over to one side. Some one-sidedness and want of symmetry are also to be found in the bones of the skull in this animal and, more or less, in all toothed whales. The nasal passage from the roof of the mouth to the external aperture, or "spiracle," may be twenty feet in length. The general colour is black, but the belly is grey. The sperm whale is a very widely diffused animal in all the warmer seas, where it may often be seen swimming with its snout raised above the surface of the water, a fact probably due to its being made buoyant by the immense mass of fat it contains. When startled it will often assume a perpendicular posture, with half the body out of the water, to look and listen. While the animal is alive, this fat is fluid, and when the whale is killed a hole is made in the outer and upper part of the head, and the liquid baled out with buckets. It solidifies on cooling, and being afterwards refined, assumes that beautifully

x

white crystallised appearance which spermaceti presents. The cachalot feeds mainly on cuttle-fishes, but also eats true fishes, even of considerable size.

The bottle-nosed whale, or hyperoödon, is a curious form which has only two teeth in the front of the lower jaw, and these are concealed in the gum. It agrees with the sperm whale in carrying a large quantity of spermaceti, yielding oil, in the upper part of its head, and blubber producing sperm oil. It attains a length of thirty feet, though females do not exceed twenty-four feet. Captain Gray tells us that these whales are occasionally met with westward, near the Shetland Isles, in March, and across the Atlantic Ocean until the ice is reached, near the margin of which they are found in the greatest numbers; but they are seldom seen among it. They are also to be met with from the entrance of Hudson's Straits and up Davis's Straits as far as 70° North latitude, and down the east side round Cape Farewell, all round Iceland, north along the Greenland ice to 77° North latitude, and also to 19° East longitude. From the fact that they are not seen in summer further south than a day's sail from the ice, it would appear that they migrate south in the autumn, and north again in the spring. They are gregarious in their habits, going in herds of from four to ten. It is rare to see more than the latter number together, although many different herds are frequently in sight at the same time. The adult males very often go by themselves, but young bulls, cows, and calves, with an old male as a leader, are sometimes seen together. They are very unsuspicious, coming close alongside a ship, round about and underneath the boats, until their curiosity is satisfied. The herd never leaves a wounded companion so long as it is alive, but they desert it immediately when dead, and if another can be harpooned before the one previously struck is killed, the hunters will often capture a whole herd, frequently taking ten, and sometimes fifteen, before the hold on them is lost. They come from every point of the compass toward the struck one in the most mysterious manner. They have great endurance, and are very difficult to kill, seldom taking less than from three hundred to four hundred fathoms of line, and stray, full-grown males will run out seven hundred fathoms, remaining under water for the long period of two hours, coming to the surface again as fresh as if they had never been away; and if they are relieved of the weight by the lines being hauled in off them before they receive a second harpoon and a well-placed lance or two, it often takes hours to kill them. They never die without a hard struggle, lashing the sea white about them, leaping out of the water, striking the boats with their tails, running against them with their heads, and sometimes staving the planks in, and freequently towing two heavy whale-boats about after them with great rapidity. The young are black, the old light brown, and the very old almost yellow. The jaws, front of the head, and a band round the neck, white. The belly greyish white. Their tails are not notched in the centre as are those of most other whales. They can leap many feet out of the water, even having time while in the air to turn round their heads and look about them, taking the water head first, and not falling helplessly into it sideways, like the larger whales. A full-grown specimen will yield two tons of oil, besides two hundred-weight of spermaceti. They live on cuttle-fishes. Certain allied species from a small group characterised by having a considerable sized tooth on either side of the lower jaw. One of these, named after Mr. Layard, has a pair which, as age advances, become very long, narrow, flat, curved teeth, like a pair of bony straps, that at last curve inward over the upper jaw, the movements of which they must hamper. These whales have the bony support of the upper jaw in the form of a long, cylindrical bone, or "rostrum," denser than ivory, and such structures, more or less mutilated, are frequently found fossil in Pliocene strata.

A most curious Arctic cetacean is the "narwhal," or sea unicorn, the latter name having been given to it on account of an enormous tusk which the males develop. The length from head to the end of the tail, without

FIG. 81.

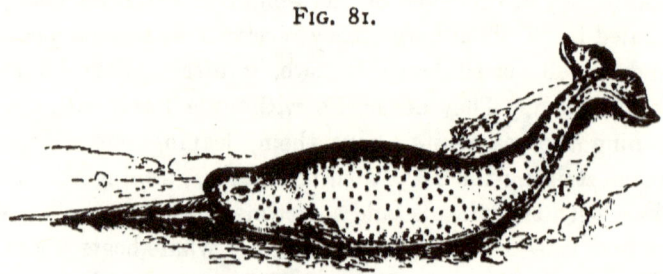

THE NARWHAL.

the tusk, is about fifteen feet, but the tusk itself often attains a length of seven or eight feet. The head is short and rounded and the paddles very broad. In colour it is dark grey above, white below, and the whole body is marbled or spotted with blackish, or more or less dark, grey. It feeds on small fishes, cuttle-fishes, and crab-like animals. It has a few irregular rudimentary teeth, but besides them two elongated teeth lie horizontally within the upper jaw in the female. In the male one of these, usually the left one, becomes enormously developed, jutting straight outward from the front of the head like a great horn. It is marked with spiral grooves and ridges, and tapers gradually to a point.

Sometimes, but rarely, both teeth are thus developed. The narwhal is seldom to be met with south of 65° North latitude, but it has at least visited the British coasts three times: once in 1678 it entered the Firth of Forth, in 1800 it came near Boston in Lincolnshire, and in 1808 another visited Shetland. In the Middle Ages it seems that the tusks of these animals were regarded as unicorns' horns, and therefore, on account of the great medical virtues attributed to them, fragments would sometimes fetch more than ten times their weight in gold. Old legends assert that the unicorn, when he goes to drink, first dips his horn in the water to purify it, and that other beasts delay to quench their thirst till the unicorn has thus sweetened the water.

Scoresby describes narwhals as extremely playful, frequently elevating their horns and crossing them with each other as in fencing, but they have never been known to strike and pierce the bottoms of ships as swordfish often do. The blubber is usually about three inches in thickness and amounts to nearly half a ton in weight.

The beluga, or white whale, is a handsome animal of the Arctic Seas and American coast, as far south, at least, as the river St. Lawrence, which it ascends to a considerable distance. It is about twelve feet long. In the year 1815 one was observed for three months swimming in the Firth of Forth. When met with in "schools" (for they are gregarious animals), they are not at all shy, but often follow ships in herds of thirty or forty, and form a remarkable sight from the beautiful white colour of the adult animals as they leap and gambol in the midst of a calm, dark sea. The flesh has been said to be fairly good eating. In the beluga and all the cetaceans which remain to be noticed, there are teeth in the upper as well as in the lower jaw.

The grampus is a very powerful and ferocious beast, which ranges the ocean from Greenland to New Zealand, and sometimes attains a length of twenty feet. It may be known when seen swimming by its dorsal fin, which is narrow and very high. Grampuses are the only cetaceans which habitually prey on warm-blooded animals. Indeed, this species is the most voracious and destructive inhabitant of the ocean. Eschricht found in a large specimen, the stomach of which measured 6 feet by 4, the bodies of various seals, flayed and so twisted that they had to be extricated one by one to count them. There were also porpoises in it, though the body of only one was entire. Altogether it contained the remains of thirteen seals and thirteen porpoises, besides one very small seal. But grampuses devour fish as well as warm-blooded animals, and among the members of their own order they will even attack whales, combining in packs to hunt down and destroy them, as wolves combine to hunt down and destroy different kinds of cattle. On the north-west coast of America grampuses have been known to seize and bear away a whale which had been captured by whalers, in spite of all their efforts to prevent them.

Many fabulous tales have been told of the grampus, and one of them relates to their practice of attacking whales. It has been said, for example, that they hunt the whale in order to gratify a somewhat refined and luxurious taste for "whale tongue." They have been said expressly to worry and harass their huge victim for the purpose of making him, in his agony, open his mouth. Then the grampus was said to dart instantly upon its tongue, seize it and tear it out, in order to enjoy so delicious a morsel.

The common porpoise is, of course, by far the best known of the British cetacea. When full-grown it attains a length of five or six feet. It may every now and then be seen in the river Thames, where it has ascended to Richmond, and it has also reached Neuilly, on the Seine. It frequents the coasts of the United States as well as those of Europe, but it rarely passes through the Straits of Gibraltar. It is very destructive of fish, feeding voraciously on mackerel, pilchard, and herrings. Such is its eagerness in pursuit of the last named that it is often caught by fishermen in their herring nets. At one time it was commonly eaten both in France and England, and was deemed a valuable addition to the table on a day of abstinence. Malcolm IV. of Scotland granted to the monastery of Dunfermline the porpoises caught in its vicinity.

As we have before mentioned, roast porpoise figured in the banquet given by King Richard II., in Westminster Hall, on the day of his coronation, which happened to be a Friday, so that no "flesh meat" could be partaken of at it. It was esteemed in England as late as the time of Queen Elizabeth, and was eaten with a sauce of breadcrumbs and vinegar. Its skin is sometimes used as leather, and is valued for its strength, while its blubber furnishes oil.

The porpoise is gregarious, and most persons who live near the coast must often have observed its gambols. On the approach of a storm, and even in the middle of one, they seem to revel in the waves, frequently showing their black backs above the surface, and often throwing themselves clean out of the water in a vigorous leap. Two which were taken in Wareham River about 1817, yielded sixteen gallons of oil. One of them was found to have milk, which, when tasted, was declared to be salt and fishy. In 1820 three more were driven right up the same river to the town of Wareham. Then a fence was put across the river, both above and below them, in order that they might be exhibited. They plunged violently, however, and their cries—which they continued during the night as well as during the day—were so distressing, that after the third day of their captivity they were taken from the water, killed, and cut up.

In the porpoise, as in the right whale, the seven neck bones unite together. The stomach is divided into three chambers, thus reminding us of some of the even-toed ungulates. The brain is very large and very broad, and is grooved over the surface in a complex pattern. The nasal passages which are, as in all air-breathing animals, double in the skull, unite and open on the convex surface of the head by a single external aperture. The canal which passes from the upper surface of the skull to the exterior, is dilated into certain chambers which have elastic and muscular walls, by which means the forcible ejection of the breath—*i.e.*, the "spouting"—is the more readily effected.

The name "porpoise" seems to be derived from the French *porc-poisson*, or the Italian *porco* and *pesce*. Its French name *marsouin*, on the other hand, corresponds with the old German word "marsuin," which is the same as the German "meerschwein," which is "sea-hog" in English.

The bay porpoise of the Pacific United States is one of the smallest of the cetacea.

The pilot-whale, or round-headed porpoise, remarkable as it is for the shape of its head, is still more remarkable for the length of its paddle, or pectoral fin. It is not that the parts answering to our upper arm and our forearm are lengthened: it is the digits which are so long, at least the second and third of them. The second, which is the longest, contains no less than twelve or thirteen bones, while the third has nine. There is nothing like this elsewhere in the whole class of beasts. Even the long fingers of the bats never have more bones than have our own fingers.

As to the head, it is very projecting and rounded in front, a fact due to the presence of a great cushion of fat on the anterior part of the upper jaw and in front of the blow-hole (or single external aperture) of the nasal

FIG. 82.

THE ROUND-HEADED PORPOISE.

passages. The animal is of a deep rich black, except the throat and belly, which are white. It is very widely distributed from the North Atlantic to Australia, and it has been so often observed that its habits are pretty well known. It will eat herrings, ling, and such creatures, but its favourite food consists of cuttle-fishes. The round-headed porpoise is a very gregarious animal, and, very unlike the grampus, is mild and inoffensive as well as sociable. Their sociability is fatal to them, since as soon as one is attacked or driven on shore they instinctively rush together and blindly follow the stranded individual till they are stranded also. The inhabitants of the Orkney, Shetland, and Faroe islands and the Icelanders, get an abundant supply of oil from them. On the appearance of a shoal, the men of the locality assemble, and the sailors try to get to seaward of them and then drive them into shore by shouts and missiles. It seems that the cries of distress of the first victims further aid in attracting others to their vicinity. In 1809 a shoal of eleven hundred were thus taken in Iceland. In 1814 a hundred and fifty were driven into Belta Sound, in Shetland, and there despatched.

The short-finned roundhead, frequents the Atlantic

FIG. 83.

THE COMMON DOLPHIN.

coast of the Middle and Southern United States, while Scammon's roundhead is found off the Pacific coast of North America, where it assembles in large "schools," and often enters bays and lagoons to feed on small fish.

The dolphin, a creature from six to eight feet long, pertains to a group of cetaceans which differs from that to which the porpoise belongs in that, instead of having a head rounded in shape, their jaws present the appearance of a long, narrow beak, like that of many birds. It is an animal renowned both in classical and mediæval literature. It was a sacred fish to the Greeks, religiously venerated because when Apollo appeared to the Cretans and obliged them to settle on the coast of Delphis, he did so under the form of a dolphin. Therefore it was that at his world-renowned oracle of Delphi he was worshipped under the symbol of that cetacean. It was also credited with a warm attachment to mankind, readily lending its aid in cases of shipwreck and disasters of various kinds. Thus Phalantus, the founder of Tarentum, when wrecked on the Italian coast, was, we are told, carried to shore by a dolphin. No doubt many readers also know Ovid's tale about the musician Arion, who, when about to be thrown overboard by sailors who coveted his possessions, begged that he might be permitted to play a last melody, which attracted admiring dolphins, one of which bore him safely back to Tanarus.

"Secure he sits, and with harmonious strains
Requites his bearer for his friendly pains."

Pliny also tells of a dolphin which daily carried a lad to and from his school, across Lake Lucrinus, in Campania, and, after the lad's death, died of a broken heart! The shield and sword of Ulysses are described as having borne the image of a dolphin, and it figures on many ancient coins—though for the most part very incorrectly —with a rounded head and spiny fins. Yet, on an ancient Syracusan coin in the British Museum the creature is very faithfully depicted. It shows a spiny back also in heraldry, in which science it is reckoned as "the king of fishes." It appears in several coats-of-arms, and, amongst others, in that of Fitz-James, which bears "*a dolphin naiant embowed*," heraldic terms, denoting it as "swimming with a curved back."

The name of dolphin, in French *Dauphin*, was also adopted as the title of the eldest son of His most Christian

Majesty, the king of France and of Navarre, it is said in the following way : The Counts of Grenoble, who in the ninth century were feudatories of the kingdom of Arles, subsequently became, as Counts of Vienne, independent, and Count Guy VIII. became surnamed " Le Dauphin " because he wore one on his helmet and shield, and his territory was called " Dauphiné." The last of that dynasty, having lost his only son, gave up his sovereignty to King Philippe of Valois in 1349 and became a friar. From that time the king's eldest son was known by this title, as the English king's was by that of Prince of Wales.

The flesh of the dolphin, like that of the porpoise, used to be eaten in Lent, and the celebrated Dr. Caius of Cambridge says that in his time it was considered a delicacy.

Its voice consists of a low murmur. It has but a single young one at a time, which the female treats with great tenderness and care. Its milk is abundant and rich.

Dolphins are very voracious and eat large quantities of fish. They often approach fishermen's nets, and they doubtless sometimes follow ships for the sake of food. They swim with great velocity, and can shoot ahead of vessels and round them although they may at the time be scudding rapidly before the breeze.

The common dolphin is found in the Mediterranean and Atlantic, while closely allied forms inhabit the Australian seas and the North Pacific.

A very curious cetacean inhabits the Ganges, Brahmaputra, and the Indus. It is found even in the smallest tributaries of those rivers, where there is water enough for it to swim, but it never passes out into the open sea. It attains eight feet in length, and is known in Bengal as the soosoo. It is blind, and gropes about with its long beak-like snout in muddy waters for the small fishes, shrimps, and crabs on which it feeds.

Another exclusively river cetacean, named *Inia*, is found in the Upper Amazon and its tributaries. It is about seven feet long, and has an elongated beak. A still longer beak is possessed by a similar creature which is found at the mouth of the Rio de la Plata, and has been named *Pontoporia*. It is about the smallest of the cetaceans, and does not exceed five feet in length, but it has from fifty to sixty teeth on either side of each jaw.

Such are the more noticeable existing forms of toothed and toothless whales. Immense quantities of allied forms have been found fossil in later tertiary strata, especially in Belgium and the east of England.

The illustrious American naturalist, Dr. Harlan, found a tooth in the Eocene strata of Alabama, to which he gave the name *Basilosaurus*, but Sir Richard Owen pronounced it to belong to a beast, which—from the form of the tooth—he named *Zeuglodon*. Herr Kock also found fossil in America, many years ago, a number of bones of the backbone of some animal. These he concocted into an immense creature 100 feet long, which he called the *Hydrarchus*. It was taken to Europe, when the great John Müller saw it at Berlin. He gave a correct description of it, showing that it was really but 60 feet long at the most, as also that its backbone was formed like that of cetaceans. It was a zeuglodon. Now we know that the skull and teeth of that animal are very seal-like, and there is much reason to believe that, altogether, this enigmatical creature was much more like a seal than it was like any kind of whale.

This consideration leads us to make a few remarks as to the origin of the cetacea. Whales, whether toothed or not toothed, have certainly nothing to do with mermaids or sirenia. The zeuglodon seems to point to a direct connection between them and the seals, but the cetacean-like structure of that creature's backbone may be merely a resemblance induced by similar habitual needs and no sign of real affinity.

That the whales, like the mermaids, have descended from some four-legged beasts, is shown by the fact of their possessing the rudiments of hind limbs, just as the rudimentary teeth of unborn whalebone whales show that such whales had animals with ordinary teeth for their ancestors, and that their wonderful "baleen" is a comparatively modern improvement. Wherever they come from they must have been evolved since the deposition of the chalk, as it is incredible that had they existed before that period, none of their remains should have been preserved in the cretacean rocks. The presence of the soosoo and of inia in rivers only, points to the possibility of all the cetacea having descended from river-inhabiting species, while Sir William Flower, who has made the whales his special study, deems it probable that they are descendants of some more or less hog-like creature. There are, indeed, many anatomical points of resemblance between the porpoise and the hog.

A word may be said in conclusion as to the wonderful brain of the porpoise and other cetaceans. It cannot evidently be a sign of the possession of intellectual faculties beyond those of other brutes. When we recall to mind the fact that the sluggish, torpid manatee has a very simple brain, it seems that the large and richly convoluted condition of the very active and vivacious porpoise, may be due to the need of a nervous supply sufficient to carry on the constantly vigorous movements of such a warm-blooded inhabitant of the ocean. This thought may serve as a caution against that hasty attribution of a direct connection between intellectual power and the development of certain superficial parts of the brain, which has been so widely diffused a belief of our own days.

## XII

## THE OTHER BEASTS

WITH the present essay we shall bring to a close this series of sketches, and will herein endeavour to collect as it were into a focus, such light as we have been able to obtain with respect to the various groups of animals which have served as our types. The information as yet conveyed has necessarily been fragmentary. It is time to present it as an orderly whole. These dozen chapters have been intended to serve as an introduction to a knowledge of zoölogy, and especially of the leading section of that primary division of animals to which we ourselves belong. That primary division is made up of all those creatures which have either a "back-bone" or a representative thereof, formed of gristle or some softer substance. As most of such creatures possess, as we do, a spinal column or back-bone, formed of a chain of small bones, each of which is termed a "vertebra," this whole primary division of animals is spoken of as the "vertebrate" division, or the division "vertebrata." This division is often also called a "sub-kingdom," because it and the other "sub-kingdoms" together include all animals, and animals taken as one great whole, have been fancifully regarded as a kingdom. The *Animal Kingdom* being thus opposed to, and contrasted with, the whole mass of plants or the *Vegetable Kingdom*. Besides the "sub-kingdom," or primary division, of "vertebrate" or "back-boned" animals, there are various other such primary divisions, as for example the sub-kingdom of molluscs or mollusca (squids, snails, oysters, &c.); the sub-kingdom of creatures with jointed legs or arthropoda (crabs, shrimps, insects, spiders, scorpions, hundred-legs, &c.); the sub-kingdom of worms or vermes; that of star-fishes or echinoderma, and so on. But with all these we have here nothing further to do; we have but to recognise the fact that our own sub-kingdom is but one of a certain number of other such primary groups.

In our essay on the opossum,* the fact was roughly indicated that the division of "back-boned animals" was made up of certain *classes*—namely, mammals, birds, reptiles, and fishes; but in that on the bull-frog, it was further pointed out that frogs, efts, and their allies, should more properly constitute a group by themselves, and so constitute a distinct class, "batrachia."

Now these five classes may be taken as falling into two sections which are very distinct, but which it will here suffice to characterise (apart from exceptional legless forms) as, (1) those having limbs made up of arm and hand, or leg and foot respectively, and (2) those the limbs of which are not so made up. The limbs of all beasts, birds, reptiles, and batrachians, are thus composed, but those of all other back-boned animals are not thus composed; therefore, they are not so composed in fishes, and so fishes are not included amongst the animals noticed in these essays.

The groups of the class of birds, have been more or less indicated in our study of the turkey,† those of the class of reptiles have been considered with the rattle-snake,‡ and those of the class batrachia in our notice of the bull-frog.§

* See p. 42.   † See p. 66.   ‡ See p. 122.   § See p. 96.

Y

But the class which is naturally most interesting to us is the class to which we ourselves belong, the class *Mammalia*. When treating of the "opossum" we showed* how the class of beasts was divisible into a number of orders which could be arranged in three distinct groups; one such group (the lowest) included only the platypus and echidna, which together constitute by themselves the order Monotremata. The second group was constituted by the pouched beasts or the placental mammals alone, which form the single order *Marsupialia*, while all other mammals which make up a variety of orders, constitute the third division of placental mammals. Of these we have drawn attention to the apes and the bats, the carnivorous beasts,† the seals and the sea-bears, the hoofed-beasts, the sirenia and cetacea,‡ and the sloths, ant-eaters, and armadillos. A few groups yet remain to be noticed, and to them we devote this final article. They are (1) the lemurs; (2) the rodents; and (3) the insect-eating beasts. Having said what we have to say about these, we shall then be in a position to survey and summarise what knowledge we may have gained about that whole class, which includes man and beast— the class Mammalia.

As to the lemurs,§ we must premise that the term may be taken in a wide or in a narrow sense. There are creatures to which that term properly and specially applies, while there are various others more or less different forms, which are also sometimes called lemurs, but which are but distant relations of the animals, to which the term specially applies. To avoid ambiguity, we shall henceforth speak of the entire group (of lemurs

* See pp. 43 and 62. † "Racoon," see p. 211.
‡ "Whales and Mermaids," see p. 303.
§ "Monkeys," see pp. 7, 33 and 34.

proper, together with their more or less distant relatives) as "lemuroids" reserving the word *lemurs* for the animals most properly so called.

FIG. 84.

THE RING-TAILED LEMUR.

Such true lemurs are animals of about the size of a cat, with sharp-pointed muzzles and long tails well clothed with hair, while a soft thick fur invests the whole body. Their legs are not much longer than their arms, while each extremity is, as in the case of the monkeys, modi-

340  TYPES OF ANIMAL LIFE

fied to serve as a hand, the great toe as well as the thumb being opposable to the other digits. The thumb is well developed, but the second or index digit is rather short.

Fig. 85.

THE SHORT-TAILED INDRIS

These animals are very common in our menageries, one of the handsomest being known as the ring-tailed lemur, because its tail is marked with alternate rings of black and white, its body being clothed with fur of a delicate

FIG. 86.

THE LONG-TAILED INDRIS.

grey colour. It is not only an extremely active and graceful animal, but also a very gentle one, making an excellent pet (Fig. 84).

All the lemurs are inhabitants exclusively of the Island of Madagascar, where they live in small troops in the woods, which they make resound with their cries; they are excellent climbers, but, when on the ground, remain on all fours. In sleeping they wrap their long tails round their bodies. Their food consists of fruits, eggs, young birds, and insects, and they seek it by day, though they are most active towards evening. They have one or two young at a birth, which at first are nearly naked, and are carried about by the mother, clinging to her belly and almost concealed by the long hair which clothes it. Lemurs have teeth much like those of monkeys, and like them and ourselves have but four cutting teeth in the middle of each jaw, though it looks at first sight as if there were six in the lower one, since the lower eye-teeth are formed like the cutting-teeth which are adjacent to them.

Madagascar also possesses a curious group of lemuroids called indris—short-tailed, long-tailed, and woolly —which have only two cutting-teeth in the lower jaw.

The short-tailed indris is the giant amongst lemuroids, its head and body together measuring two feet. It lives in the forests of the eastern part of Madagascar, going about in small parties of four or five individuals. Its hind legs are much longer than its fore limbs, and its great toes are very large, and on the ground it assumes an upright attitude (Fig. 85).

Of long-tailed indris there are at least three different kinds, and one or other of these species is to be found all over Madagascar, living in small troops of six or eight. They are also large animals which are very arboreal, leaping from tree to tree without apparent effort, sometimes for a distance of ten yards. On the ground they go erect, but effect their progression there in a very ludicrous fashion. Standing upright, they hold their arms over their heads, and then make a series of short jumps; they are gentle animals, but somewhat stupid. They sleep at night and during the heat of the day, but are active morning and evening. Their food consists mainly of birds, flowers, and fruits (Fig. 86).

Certain small nocturnal lemuroids also inhabit Madagascar. They are of small size, some being smaller than a rat. Their most interesting structural peculiarity, of a permanent nature, lies in their ankles. Instead of these parts consisting (as they do in ourselves, monkeys, and almost all other beasts) of a group of short bones only, two of them are elongated and lie side by side, so adding an additional segment to the limb—one intermediate between the leg and the foot. They have also another interesting peculiarity of a temporary nature, this is their tendency to accumulate a quantity of fat in certain parts of the body, especially at the root of the tail, which becomes of an exceedingly large size. This peculiarity of structure is related to a peculiarity of habit: for during the dry season they retire into the holes of trees, coil themselves up and pass the whole period in sleep, as bats with us hibernate in winter. When, with the advent of the rainy season, they rouse themselves again, their fat has disappeared: having served to nourish them during their period of torpor.

All the lemuroids yet noted are inhabitants of Madagascar, but another group, called galagos, are none of them found within that island, but all of them are exclusively inhabitants of Africa south of the Sahara, they are active only at night when they feed on fruits,

344  TYPES OF ANIMAL LIFE

insects, eggs, and small birds. Their size varies from that of a small rabbit to that of a large mouse. They

Fig. 87.

THE SENEGAL GALAGO.

THE OTHER BEASTS 345

have very large eyes and ears, with long tails, and are
clothed in soft woolly fur. Habitually they live in trees,
but when they descend to the ground they progress by
long jumps as kangaroos do, their power to effect these

FIG. 88.

THE SLENDER LORIS.

depends on the fact that two of their ankle bones are
even more elongated than in the small Madagascar
lemuroids last noted.

A most singular lemuroid is the slender loris—a creature with no trace of a tail, and limbs which have much
the proportions of our own as to length, though they

are extremely attenuated. The animal therefore looks like a very small dwarf (about the size of a squirrel) reduced almost to a skeleton. It inhabits only Southern India and Ceylon. Its index finger is exceedingly short, but its eyes are very large indeed. It is regarded by the natives of India as a remedy for ophthalmia, on which account it constitutes an article of commerce in the bazaars of Madras. It is a slow-moving creature, strictly nocturnal, and feeding on young shoots and leaves, insects, lizards, eggs, and small birds. It is also said to be extremely fond of honey.

Another eastern lemuroid is the slow loris, which ranges from Cochin China to Sumatra and Borneo. It is generally like the slender loris, except that it is much stouter. Excessively slow in its movements and sleeping by day, it creeps about at night in the forests it inhabits, never jumping or running, but always clinging with great tenacity to the branches. The arrangement of the tendons of its feet is such that the mere weight of its body suffices to keep its toes strongly bent, and firmly clasping any object from which it may hang. Its food consists of fruit, young shoots, insects, and birds, and it will seize such of the last as its noiseless approach at night may enable it to surprise when they are at roost.

Two allied kinds exist in Africa, one of these, the angwantibo, inhabits Old Calabar. The other is the potto, which is interesting as being one of the first lemuroids discovered. One, Bosman, a traveller, met with it during his voyage to Guinea, and described it in the year 1705. After that, it was never again seen by a European for twenty years, and it was only first fully described in 1830. Both these African forms are very like the slow loris, and resemble it in their habits, but they both have

## THE OTHER BEASTS

the index finger reduced to a mere rudiment. It is especially rudimentary in the potto; and this forms no slight argument against Darwinism, that is, against the doctrine that species have been formed by natural selec-

FIG. 89.

THE POTTO.

tion. For how is it possible that the potto's life could have been repeatedly saved on account of its *not* possessing an index finger? The Indian Archipelago, that is to say, Sumatra, Borneo, Celebes, and the Philippine Islands, possess a singular little animal, not so big as the common squirrel, which is a great contrast to the slow-moving lemuroids last noticed. This little animal is the tarsier. To a certain extent it resembles the galagos. It has large eyes and ears, and a long tail tufted at the end, but it carries to a yet greater degree

FIG. 90.

THE TARSIER.

the prolongation of the two ankle bones, so that there is a most distinct thin elongated segment of the leg between the shin and the foot. This gives the little animal great power of jumping, and in that way it moves about actively at night, in pursuit of insects and other small animals on which it largely feeds. It has but two cutting teeth in the lower jaw, and though there are four in the upper one, the two middle ones are large and closely applied one against the other.

The last lemuroid to be here noticed is a still

FIG. 91.

THE AYE-AYE.

more exceptional one. It is called the aye-aye, and is found nowhere but in Madagascar. There it was discovered by Soumerat in 1780, and was described in Buffon's "Natural History" from a skin which Sonnerat presented to the King's Cabinet. That skin has survived all the tumults of the Revolution, and it constituted the only evidence of the existence of such a beast, for more than half a century. At last, in 1844, a second specimen was brought to Paris, and this was followed by others, and live ones were then obtained for the Zoölogical Gardens in London, where one may now be seen. The aye-aye is about the size of a cat, and has long dark hair, a long tail, a round head, and very large ears. The hind foot is like that of the lemuroids, but its hands have long fingers with pointed claws, and the middle finger of each hand is long, out of all proportion and more slender than the others.

It is a strictly nocturnal animal, passing the day in a round nest formed of leaves and fixed between the forked branches of some tree. Its food has not been certainly ascertained. It has been said to feed merely on grubs which live in the wood of trees. Its large ears have been represented as enabling it to hear the grubs there at work, while its large teeth could gnaw the wood quickly and so lay them bare. Then as such grubs retreated deeper from the light of day, it was said to introduce its long and slender middle finger into their burrow and hook them out. But observations made on aye-ayes in confinement do not confirm this representation, but rather point to their supporting themselves on the succulent juices of plants. Their teeth certainly seem to indicate a vegetable diet, and they are formed like those of the animals next to be here noticed, namely the rodents. There are but two cutting teeth in either jaw, but these are very large, and those of each jaw applied closely, one against the other; they are, moreover, separated from the grinding teeth by a long toothless space, there being no eye-teeth at all, though such exist in each of the jaws of all the other lemuroids.

On account of these teeth, the true nature of this creature was long misunderstood. It was taken to be really one of the gnawing animals, or rodents, and was so classed even by the great Cuvier himself. But what are lemurs? As we before pointed out,* they have no essential relationship to the apes, though they have long been classed with them, and though it is convenient that they should so remain, for they have no known more essential relationship to any other group of beasts. Some tertiary fossils have indeed been supposed to indicate a connection between them and hoofed quadrupeds, but we regard this as being most problematical. The resemblance, which exists between our own species and the group of apes, was insisted on in the very commencement of the first of these essays, and since zoölogical classification reposes exclusively upon characteristics of bodily structure, man, when merely considered zoölogically, must be taken as a member of that order to which the apes belong. The difference between him and them, in this respect, is small indeed, when compared with that which exists between apes and lemuroids. On that account, if both lemuroids and apes are to be classed in one order, such order must be divided into two suborders, in one of which will stand man and the apes, while the lemuroids must occupy the other.

Of the gnawing animals, or Rodents, that well-known American animal, the beaver, may stand for us as a type. It is in much danger of extinction through the spread of population, though the cessation of the fashion of beaver-hats must tell somewhat in its favour. In the early days of this century, when that fashion was in vogue, some

* See ante, p. 34.

200,000 skins were exported annually to Europe. The European variety has become almost extinct, save where it is protected by the Emperor of Austria; though isolated pairs are met with in Germany and Russia. It was once an inhabitant of Great Britain, as some Welsh names of places would alone suffice to prove, were there not also positive testimony of it, such as that of Giraldus Cambrensis.

The European beaver never makes dams like those for which its American cousin is so renowned, yet we know, from Albertus Magnus, that it did so in the thirteenth century. The American kind was recently imported into the Island of Bute by its owner (the Marquess of Bute), and there they throve wonderfully, and formed their dams in true American fashion.

The amazing facility the beaver possesses for felling trees is due to the power of its jaws and teeth. Of these there are, as in the aye-aye, two large cutting teeth above and two below, separated by a toothless interspace from the grinding teeth behind them. Each cutting tooth is protected in front by a coating of very dense enamel, so that at its summit it wears away less quickly in front than behind, and thus a sharp, chisel-like cutting edge is constantly preserved.

Another well-known and renowned American rodent is the prairie marmot, commonly called the "prairie dog." It is a stoutly built little animal with a short tail, of very social habits, feeding on buffalo-grass, and living in large communities in burrows, so closely placed that it is very dangerous work riding across the plains they inhabit. Their habitations are peacefully shared by a small owl, and less peacefully by rattle-snakes, which latter doubtless frequent them in order to feed on their young.

THE OTHER BEASTS 353

Closely allied to it are the European marmots of the Alps, which used to be carried about by itinerant Swiss boys. As to them we learn from Professor Blasius that " they live high up in the snowy regions of the mountains, generally preferring exposed cliffs, whence they may have a clear view of any approaching danger, for which, while

FIG. 92.

THE PRAIRIE DOG.

quietly basking in the sun, or actively running about in search of food, a constant watch is kept. When one of them raises the cry of warning, a loud piercing whistle well known to travellers in the Alps, they all instantly take to flight, and hide themselves in holes and crannies among the rocks, often not re-appearing at the entrance of their hiding-place until several hours

z

have elapsed, and then frequently standing motionless on the look-out for a still longer period. Their food consists of the roots and leaves of various Alpine plants, which, like squirrels, they lift to their mouths with their fore-paws. For their winter quarters they make a large round burrow, with but one entrance, and ending in a sleeping place thickly lined with hay. Here from ten to fifteen marmots will often pass the winter, all lying closely packed together, fast asleep, until the spring."

The marmots are first cousins to the rodents most remarkable for their attractiveness. We recollect that one day when we were in the smoking-room of the Hôtel d'Angleterre at Rome, a discussion took place as to what was the most beautiful object in Nature. There were several exclamations of "a woman! a woman!" But a gentleman from Chicago said, with much deliberation, "Wal, I should say a squirrel!" Squirrels are the animals we now refer to, and they are to be found in all the warm and temperate regions of the globe, save Australia and Madagascar. They are most abundant in the Malayan region, where is to be found the giant of the group—the two-coloured squirrel—which is almost as large as a cat. In Borneo, on the other hand, there is one as small as a mouse. The European common squirrel ranges from Ireland to Japan, and from Italy to Lapland. There are altogether about seventy-five species of true squirrels, whereof fifteen are found in America. Flying squirrels, which are creatures formed like flying phalangers,* have the skin of the sides and flanks of the body extensible, and so capable of acting like a parachute. There are more than twenty-five kinds of them, yet only one species—but a most charming little one—is found in North America.

* See *ante*, p. 46.

## THE OTHER BEASTS

In West and Central Africa several species of beasts have been found which closely resemble flying squirrels, but have a number of large overlapping horny scales placed beneath the tail. On this account the name *Anomalurus* has been given them.

Those beautiful little animals, the dormice, come near the squirrels. They are unrepresented in America, but have existed in Europe from the time of the Upper Eocene. There also may be mentioned a curious animal from West Africa called *Lophiomys*, which has, from its coloration, somewhat the appearance of a skunk. It is remarkable for having a great toe which can be opposed to the digits of the foot, and for having an outgrowth from the bones of the head, extending over the side of the face (beneath the skin) and forming a sort of double-walled skull, there, as in some frogs.*

The very numerous family of rats and mice may be exemplified by that very beautiful little European animal the harvest mouse, the head and body of which scarcely exceed $2\frac{1}{2}$ inches in length. It is very elegant in shape, of a reddish brown colour above, and the under parts pure white. It builds a round nest, about the size of a cricket ball, often attached to the stalks of wheat, and formed of dry grass.

The ancient common rat of England, the black rat, has been almost exterminated by the brown rat, which seems to have come into England in the sixteenth century, the mouse and rat genus is the richest in species of any mammalian genus.

There are one hundred and thirty different kinds, some or other of which are found in all parts of the Old World save Madagascar. There are none, however, which are naturally inhabitants of America. There, nevertheless,

* See *ante*, p. 121.

is to be found the musk-rat or musquash, the head and body of which are together a foot long. Its feet are webbed, and it lives in lakes and rivers from the Rio Grande to the shores of the Arctic Seas. America also is the home of the elegant little creatures with small cheek-pouches—the vesper mice—such as the deer-mouse the wood-mouse, the golden mouse. These in the West will come and take up their abode in houses.

Gerbils are elegant creatures with large bright eyes and long tufted tails and very long hind legs, wherewith they can jump in kangaroo fashion. They range over the Old World. In the hamster of the Old World there are large cheek-pouches.

The voles are Old World creatures, which differ from rats and mice by their blunt muzzle, heavier build and less graceful movements; also by their smaller eyes and ears, and shorter limbs and tail. An allied form, *Hydromys*, exists in Australia, New Guinea and Tasmania.

The lemming of Norway is a sort of vole, very celebrated on account of its sudden and marvellous migrations. When a conjuncture of favourable circumstances enables them to multiply to an enormous extent, a migratory instinct becomes developed in them, whereby they are led to descend to lower-lying lands than those they normally frequent.

They migrate slowly and intermittently, journeying only by night, and increasing frequently as they go. Their journey may last for three years before they reach the sea coast, according to the route they may happen to have followed. When they reach the coast, they go on into the sea and so perish. As they journey along, they are preyed upon by bears, wolves, foxes, dogs, wild-cats, weasels, eagles, hawks and owls. They are also destroyed by man, and even domestic animals, such as goats and reindeer will spring upon and kill them. Numbers also die of disease, but they never turn back, they proceed ever onwards to their ultimate destination.

In Russia and Asia there are two forms of mouse-like animals, specially modified for a subterranean life, having rudimentary external ears, and a short tail, and the claws of the fore-limbs greatly developed. In Southeastern Europe there is a still more completely sub-

FIG. 93.

THE MOLE-RAT.

terranean rodent — the mole-rat, *Spalax*, which is further modified by having its eyes covered by the skin. An African mole-rat is called *Bathyergus*. Less completely subterranean is the American pouched-rat or "pocket-gopher," which lives and burrows in the plains of the Mississippi, while allied fossorial forms are found in Canada, and America west of the Rocky Mountains.

The jerboas are creatures with a kangaroo-like habit of progression. Their legs are very long, and the three middle-foot bones, which support their three toes, are anchylosed together into a sort of cannon-bone, as are

358  TYPES OF ANIMAL LIFE

the two middle-foot bones of the feet of most ruminants.*
They are all Old World forms.

A noted South American rodent is the coypu, which attains a length of two feet. It inhabits the rivers in South America and feeds on aquatic plants.

FIG. 94.

THE JERBOA.

The true "fretful porcupine" is an exclusively Old World form of rodent, which may be eaten at the Falcone Tavern at Rome. It extends south-eastwards to Borneo. In India it forms extensive burrows, often living socially and inflicting great damage on various crops. They never issue forth till after dark, but some

* See "Bison," *ante*, p. 196.

THE OTHER BEASTS 359

times do not return home before sunrise. They will actively defend themselves, charging at their foes, and at the same time erecting their spines, so that dogs are often very severely injured, or even killed by them. There are some eight different species, and there are also four

Fig. 95.

THE TRI-COLOURED TREE-PORCUPINE.

species of brush-tailed porcupines, each having a long tail with a tuft of spines towards its tip.

In America, from Canada to Southern Brazil, we also meet with porcupines of a different kind. In Brazil there are true porcupines with a long prehensile tail in harmony with their arboreal habits.

The Andes of Peru and Chili possess a rodent which may be said, as far as its exterior is concerned, to be the very opposite of a porcupine, for its fur is the softest and most delicate known. This is the chinchilla, an animal about ten inches long, with a tail half that length.

The agoutis, the paca, and the cavies, amongst which is the guinea-pig, are also exclusively American species. Only American also are two more noteworthy forms. One of these is the so-called Patagonian cavy, which is rather larger than a hare, an animal to which it bears a decided resemblance from its long hind legs and mode of progression.

The other form is the capybara, the giant of the rodent order. It is a stout bulky animal about 4 feet in length, and covered with coarse hair. These animals are to be met with on the borders of rivers and lakes in South America. They hide themselves amongst reeds and other aquatic plants, and have little to fear save from the jaguars, which are said habitually to prey upon them, as also do the pumas.

We will conclude this sketch of the great group of gnawing animals by a notice of the hare and its allies. These animals differ from all other rodents in having in the upper jaw a second and small pair of cutting teeth placed directly behind the large ones.

Of hares and rabbits there are about twenty species spread over the northern hemisphere, one of them extending downwards to South America.

The common hare is found all over Europe except Ireland, Norway and Sweden, and Northern Russia. There it is replaced by the mountain hare, which, except in Ireland, becomes snow white in winter, save the black tips of its ears. The rabbit affords one of the best—and worst—examples of the rapid diffusion of a species in regions new to it, as in New Zealand, Australia and the Falkland Islands.

Closely allied to the hares and rabbits are creatures called picas, or tailless hares, of which there are about a dozen species. They live in holes amongst the rocks of the mountains of Northern Asia, those of the Rocky Mountains of America, and, one species, in South-eastern Europe. They are small animals, which are agile and shy, and have somewhat the appearance of guinea pigs.

We must next pass on to the consideration of the order of insect-eating beasts (Insectivora)—the only order which we have now left unnoticed. It includes the moles, shrews, and hedgehogs, with other forms less familiarly known. They all have teeth with sharp points well adapted for piercing the bodies of insects and very unlike those of rodents. In the spalax and some other gnawing animals just noticed, we have met with rodents specially modified for burrowing, with very strong claws to their fore-feet, without external ears, and with eyes covered by the skin of the head. All these characters exist to the fullest degree in the true moles, which are the animals the most perfectly adapted to such a mode of life. They are confined to Europe and Asia, and there are some eight species of them. The common mole of England is found from that island to Japan, and down to the Himalayas. Its fore-limbs, with their claws, are exceedingly powerful, and moved by powerful muscles to give greater scope, for the origin of which the breast bone is keeled, as in the armadillo.* The tail is short and the body covered with a thick but short velvety fur. The mole feeds on earthworms and is most voracious. In captivity, it will eat any flesh or attack animals as big as itself. If two moles are confined together and have nothing to eat, the

* See *ante*, p. 258.

stronger one will eat the other. The mole can swim well, and burrow so rapidly in the earth, it may almost be said to fly through it. It makes a nest lined with

FIG. 96.

THE RUSSIAN DESMAN.

dry grass or leaves, from which a regular system of passages issue forth.

In America, representatives of the Old World moles are called star-moles, because a series of delicate processes of skin radiate from the extremity of the muzzle. The general form of a star-mole is like the true mole, but its tail is longer and its hand less powerful. It makes tunnels in the ground like the mole of Europe. Three other species in the United States have the hind feet webbed. They form the genus *Scalops*.

Two curious aquatic more or less mole-like creatures, but with long scaly tails, webbed feet, and long proboscis-like snouts, are known as desmans. The larger species, sixteen inches long, inhabits the lakes and streams of Southern Russia. In ancient times it was found in

FIG. 97.

THE GYMNURA.

England. The other much smaller species is found in the region of the Pyrenees. Two allied terrestrial forms, burrowing and without webbed feet, come one from Japan, the other from North America.

The shrews are a very numerous group of long-muzzled, pointed-nosed, in external appearance mouse-like creatures, which include amongst them the absolutely smallest of all beasts. Some or other species are found almost all over the world, except Australia.

The hedgehogs, animals so familiar in Europe, are animals clothed with sharp spines, which stand out defensively on every side, when the creature rolls itself

364  TYPES OF ANIMAL LIFE

up in a ball, as it invariably does when it finds itself in danger. There are about twenty different kinds of hedgehogs distributed throughout Europe, Africa, Northern Asia and Hindostan. They feed on insects,

FIG. 98.

THE DWARF TUPAIA.

slugs, mice, lizards, and snakes, and will also eat eggs, fruit, and roots. In cold countries they hibernate, that is, they pass their winter in sleep.

An allied form, the hairs of which are not spiny, though coarse, is the gymnura (Fig. 97). It is of the size of a very large rat, and has a long projecting and movable snout. It comes from the Indian Archipelago and the Malay Peninsula.

The same region is tenanted by some very elegant

FIG. 99.

THE TYPICAL JUMPING SHREW.

insectivores, known as tupaias or tree-shrews. As their name implies they are arboreal animals, and very active in their movements. They have long, more or less bushy tails, and were it not for their long pointed snouts, would greatly resemble squirrels, and they feed like those animals, sitting on their haunches, and holding their food in their fore-paws.

As these creatures resemble squirrels, so do another set of African insectivores resemble jerboas, and other

FIG. 100.

THE POTOMOGALE.

long-legged, jumping rodents. They are known as the jumping-shrews (Fig. 99). They have large ears, very long-pointed proboscis-like snouts, long tails, and very long legs

FIG. 101.

THE TAILLESS TANREC.

and feet. There are eleven different species in the group. Another African insectivore may be compared with the coypu amongst rodents. This is the aquatic form which was discovered by M. du Chaillu in Western Africa, and

THE OTHER BEASTS 367

has received the name *Potomogale*. It is about two feet in length, and has a powerful tail by which it swims. An allied form, which is small and mouse-like, inhabits the island of Madagascar. The same island possesses

FIG. 102.

THE SOLENODON OF CUBA.

some curious insectivores, which, on account of their spines, were at first taken for hedgehogs of some kind. Their type is known as *Centetes*, and is an animal from twelve to sixteen inches long. It seems to be the most prolific of all animals, for it has been said to bear twenty-one at a birth.

The islands of Hayti and Cuba possess each a species of a very curious form of insectivore called *Solenodon*.

The body is clothed with coarse hair and the fore-paws have very long claws. There is an exceedingly prolonged proboscis-like snout, and a long and naked tail.

In South Africa certain burrowing animals are found which are known as golden moles. Their eyes are covered by the hairy skin of the head, and their very small ears are concealed in their fur, the tail is rudimentary, and the claws of their fore-paws are very long and powerful. There can be no doubt but that golden moles and true moles have been evolved independently, just as

FIG. 103.

THE GOLDEN MOLE.

there seems to be no essential generic relationship between some of the burrowing rodents.

The foregoing animals are by universal consent classed together in a single order—Insectivora.

The animal before spoken of * as having recently been discovered in Australia, and called *Notoryctes*, is very interesting, because it so much resembles a golden mole, thus constituting one more striking example of the independent origin of similar structures.

We have before mentioned * the colugo, the so-called

\* See *ante*, p. 60.  † See *ante*, p. 175.

## THE OTHER BEASTS

"flying-lemur" or *Galeopithecus*. It is now generally placed, as we have placed it, in this same order. This is done, however, much less from any positive resemblance it possesses to other insectivores than from the difficulty of knowing where else to put it, and a disinclination to rank it as forming an entire order by itself. At one time it was placed with the bats, as it was before placed with the lemurs. But it certainly has no true affinity with either.

As to this species, Mr. Alfred R. Wallace tells us : *

"Another curious animal, which I met with in Singapore and in Borneo, but was more abundant in Sumatra, is galeopithecus—flying lemur. The creature has a broad membrane extending all round its body to the extremities of the toes, and to the point of the rather long tail. This enables it to pass obliquely through the air from one tree to another. It is sluggish in its motions, at least by day, going up a tree by short runs of a few feet, and then stopping a moment as if the action was difficult. It rests during the day, clinging to the trunks of trees, where its olive or brown fur, mottled with irregular whitish spots and blotches, resembles closely the colour of the mottled bark, and no doubt helps to protect it. Once, in a bright twilight, I saw one of these animals run up a trunk in a rather open place, and then glide obliquely through the air to another tree, on which it alighted near its base, and immediately began to ascend. I paced the distance from the one tree to the other, and found it to be seventy yards; and the amount of descent I estimated at not more than thirty-five or forty feet, or less than one in five. This I think proves that the animal must have some power of guiding itself through the air, otherwise in so long a distance, it would have little chance of alighting exactly upon the trunk. It feeds chiefly on leaves. The brain is very small and it possesses such remarkable tenacity of life, that it is exceedingly difficult to kill it by any ordinary means.

* "Malay Archipelago," 2nd edition, vol. i. p. 135.

The tail is prehensile, and is probably made use of as an additional support while feeding. It is said to have only a single young one at a time, and my own observation confirms this statement, for I once shot a female with a very small blind and naked little creature clinging closely to its breast, which was very much wrinkled, reminding me of the young marsupials. The fur of the back and limbs is short but exquisitely soft."

With the colugo we close our notice of the Insectivora and of all existing beasts, but before we conclude a word must be said with respect to some fossil forms. The Eocene rocks of North America and Europe have disclosed relics of the jaws and teeth of creatures which show signs of relationship to the insectivora on the one hand, and to the pouched beasts or marsupials on the other. The same relationships are also suggested by the jaws and teeth of small creatures, which have been found in secondary rocks, the jurassic formation of the United States and of England. Mammalian remains have also been quite recently discovered in the chalk.

It is time now to present in a tabular form an enumeration of all the ordinal groups of existing mammals. They may be set down as follows:

### CLASS MAMMALIA.

SUB-CLASS I.—PLACENTALIA.

| | | | |
|---|---|---|---|
| Order 1. Primates { | Sub-order A. | | Man and apes. |
| | ,, B. | | Lemuroids. |
| ,, 2. Cheiroptera | . | . . | Bats. |
| ,, 3. Insectivora | . | . . | Moles, shrews, hedgehogs, &c. |
| ,, 4. Carnivora | . | . | Racoons, bears, weasels, otters, cats, civets, and dogs. |
| ,, 5. Pinnipedia | . | . . | Sea-bears and seals. |

## THE OTHER BEASTS

SUB-CLASS I.—PLACENTALIA—*continued.*

| | | |
|---|---|---|
| Order 6. Rodentia | . . . | Squirrels, rats, hedgehogs, hares, &c. |
| „ 7. Ungulata Sub-order A. | | Bisons, antelopes, deer, camels, rhinoceroses, horses, &c. |
| | B. | Hyrax. |
| „ 8. Proboscidea | . . . | Elephant. |
| „ 9. Sirenia | . . . | Dugong and manatee. |
| „ 10. Cetacea | . . . | Whales and porpoises. |
| „ 11. Edentata | . . . | Sloths, anteaters, armadillos, pangolins. |

SUB-CLASS II.—DIDELPHIA.

| | | |
|---|---|---|
| Order 12. Marsupialia | . . . | Opossums, kangaroos, phalangers, &c. |

SUB-CLASS III.—ORNITHODELPHIA.

| | | |
|---|---|---|
| Order 13. Monotremata | . . . | Platypus and echidna. |

As to the mode of succession in which these various orders may have been evolved we can as yet only make more or less plausible conjectures.

On the whole it seems probable that the Insectivora—especially such a form as that spineless hedgehog, gymnura—may amongst existing animals give us the best general idea of primitive mammalian life.

Certain marsupials seem closely allied to insectivores, and may have been a lateral offshoot from ancestral insectivorous forms.

The platypus and echidna show characters which point down below the whole class of beasts and towards reptiles, but they are animals of very peculiar formation, and there seem to be no other existing beasts which have any special relationship to them, save perhaps some of the edentates. But these again are specially modified forms, and our speculations are here, as in so many other instances, checked by the probability of the independent origins of similar structures.

It is difficult, if not impossible, to believe that the armadillos, pangolins, and aard-vark ever had any common ancestors, save such as were near the base of the whole mammalian tree of life. The whales and porpoises on the one hand, and the sirenia (or mermaids) on the other, are both isolated groups; though we may in imagination connect the latter with the elephant and odd-toed ungulates, and the former with the non-ruminating, even-toed, hoofed beasts.

The rodents are also isolated, but that their peculiar kind of dentition has more than once arisen independently is proved to us by the wombat and the aye-aye, both of which, though no rodents, have rodent-like cutting teeth.

The seals and sea-bears or pinnipedia are doubtless modified carnivora of one kind or another, and the carnivora themselves may have been modified from early insectivora. The origin of lemuroids is, as recently stated, problematical, while as to that of bats we have as yet no fragment of evidence.

Monkeys, as we pointed out in our first article, stand alone on a veritable zoölogical island, save that the human form very closely resembles them.

As to the origin of the whole class Mammalia, we have not yet enough evidence to enable us to affirm anything with certainty, but probability points to its derivation from one or other of the earlier forms of reptilian life.

It now remains but to glance over the earth's surface and see what are the beasts which characterise, respectively, its several geographical regions. The first region, which is made up by Europe, Africa north of the Sahara, and Asia north of the Himalaya—excluding the south of Arabia—may be called the *Northern Old World region*. This is the special region of sheep, and goats, and deer, the musk-ox, the camel, the musk-deer, moles, desmans, dormice, and various special carnivores.

The second region may be called the *Ethiopic*, consisting as it does of Africa south of the Sahara, with Madagascar and Southern Arabia. This is (in South Africa) the great region of antelopes; and here the giraffe and hippopotamus now have their only home. Buffaloes and chevrotains inhabit it, also zebras, the aard-wolf, and the aard-vark, the gorilla and chimpanzee, all the baboons, the thumbless apes called *colobi*, the potto and the galagos, and the golden-moles. In Madagascar are the true lemurs, the indris, and that most peculiar beast, the aye-aye.

The third geographical region is the *Indian*, and includes Asia south of the Himalaya down to the Island of Bali. In this region we have the orang (Borneo and Sumatra). All the long-armed apes, the macaques, the slender and slow loris, the tarsier, tupaias, and the colugo, the Malayan tapir, the nilghai, the tiger, and hunting leopard, the sloth bear, chevrotains, and the four-horned antelopes.

The fourth geographical region is the *Australian*, and includes the Indian Archipelago from the Island of Lombock, Moluccas, New Guinea, Australia, Tasmania, and New Zealand. Here we find all the marsupials exclusively, save the true opossums; the monotremes, and the dingo.

The fifth region is that of *Northern New World*, and embraces North America down to North Mexico. Here alone are to be found the wapiti, the American bison, the prong-buck, and the mountain sheep, the racoon and coati, pouched-rats and vesper-mice, the musquash, the skunk, and the American badger.

Lastly we have the region of the *New World Tropics*,

which consists of America south of North Mexico, and including the Antilles. This is the region of sloths, ant-eaters and armadillos, of howling and spider monkeys, sakis, squirrel monkeys, and marmosets. This is also the region of the true vampire bat, *Desmodus*. Here also alone are found chinchillas, cavies, and the Patagonian hare, the capybara, or the jaguar and the ocelots; and lastly, here are found the tapirs, other than that of the Malay Archipelago.

We have incidentally noted that various forms are absent from the Antilles and Madagascar, which we might have expected to find in countries so warm, and with such abundant vegetation. More noteworthy still is it that all Oceanic Islands are devoid of beasts, save bats and such as might have been imported by man or have been accidentally transported by floating timber, &c.

It is evident that the distribution of animals over the earth's surface to-day, or their distribution through past time, as evidenced by their fossil remains, both point to a gradual and natural origin and distribution of every kind of beast composing the mammalian class. We say of every kind of *beast*, because as regards man, no reasonable opinion could be gathered from the facts set down in this series of essays. The great distinguishing characteristic of man is his intellectual energy, above all, his power of perceiving that a difference exists between right and wrong, between duty and pleasure. But no inquiry as to the human mind has been here attempted, for the only purpose of the present work is to serve as an introduction to the study of the higher animals, especially those which constitute the class of beasts—the class Mammalia.

*Printed by* BALLANTYNE, HANSON & CO.
*London and Edinburgh*

www.ingramcontent.com/pod-product-compliance
Lightning Source LLC
Chambersburg PA
CBHW030400230426
43664CB00007BB/670